Participatory approaches to the conservation and use of plant genetic resources

Esbern Friis-Hansen and Bhuwon Sthapit, *editors*

The International Plant Genetic Resources Institute (IPGRI) is an autonomous international scientific organization, supported by the Consultative Group on International Agricultural Research (CGIAR). IPGRI's mandate is to advance the conservation and use of genetic diversity for the well-being of present and future generations. IPGRI's headquarters is based in Rome, Italy, with offices in another 19 countries worldwide. It operates through three programmes: (1) the Plant Genetic Resources Programme, (2) the CGIAR Genetic Resources Support Programme, and (3) the International Network for the Improvement of Banana and Plantain (INIBAP).

The international status of IPGRI is conferred under an Establishment Agreement which, by January 2000, had been signed and ratified by the Governments of Algeria, Australia, Belgium, Benin, Bolivia, Brazil, Burkina Faso, Cameroon, Chile, China, Congo, Costa Rica, Côte d'Ivoire, Cyprus, Czech Republic, Denmark, Ecuador, Egypt, Greece, Guinea, Hungary, India, Indonesia, Iran, Israel, Italy, Jordan, Kenya, Malaysia, Mauritania, Morocco, Norway, Pakistan, Panama, Peru, Poland, Portugal, Romania, Russia, Senegal, Slovakia, Sudan, Switzerland, Syria, Tunisia, Turkey, Uganda and Ukraine.

Financial support for the Research Agenda of IPGRI is provided by the Governments of Australia, Austria, Belgium, Brazil, Bulgaria, Canada, China, Croatia, Cyprus, Czech Republic, Denmark, Estonia, F.R. Yugoslavia (Serbia and Montenegro), Finland, France, Germany, Greece, Hungary, Iceland, India, Ireland, Israel, Italy, Japan, Republic of Korea, Latvia, Lithuania, Luxembourg, Macedonia (F.Y.R.), Malta, Mexico, the Netherlands, Norway, Peru, the Philippines, Poland, Portugal, Romania, Slovakia, Slovenia, South Africa, Spain, Sweden, Switzerland, Turkey, the UK, the USA and by the Asian Development Bank, Common Fund for Commodities, Technical Centre for Agricultural and Rural Cooperation (CTA), European Environment Agency (EEA), European Union, Food and Agriculture Organization of the United Nations (FAO), International Development Research Centre (IDRC), International Fund for Agricultural Development (IFAD), Interamerican Development Bank, Natural Resources Institute (NRI), Centre de coopération internationale en recherche agronomique pour le développement (CIRAD), Nordic Genebank, Rockefeller Foundation, United Nations Development Programme (UNDP), United Nations Environment Programme (UNEP), TBRI and the World Bank.

The geographical designations employed and the presentation of material in this publication do not imply the expression of any opinion whatsoever on the part of IPGRI or the CGIAR concerning the legal status of any country, territory, city or area or its authorities, or concerning the delimitation of its frontiers or boundaries. Similarly, the views expressed are those of the authors and do not necessarily reflect the views of these participating organizations.

Citation:
Friis-Hansen, Esbern and Bhuwon Sthapit, editors. 2000. Participatory approaches to the conservation and use of plant genetic resources. International Plant Genetic Resources Institute, Rome, Italy.

ISBN 92-9043-444-9

IPGRI
Via delle Sette Chiese, 142
00145 Rome
Italy

© International Plant Genetic Resources Institute, 2000

Contents

Preface 5

Acknowledgements 6

Contributors 7

Introduction 11
       Esbern Friis-Hansen and Bhuwon Sthapit

Section I. Crosscutting issues

1. Concepts and rationale of participatory approaches to conservation and
   use of plant genetic resources 16
   Esbern Friis-Hansen and Bhuwon Sthapit

2. Institutional perspectives on participatory approaches to use and conservation
   of agrobiodiversity 22
   Conny Almekinders and Walter de Boef

3. A brief review of participatory tools and techniques for the conservation and
   use of plant genetic resources 27
   Amanda King

4. Integrating gender analysis for participatory genetic resources management:
   technical relevance, equity and impact 44
   Maria Fernandez, Pratap Shrestha and Pablo Eyzaguirre

Section II. Enhancing farmers' access to plant genetic resources maintained *ex situ*

5. Overview 54
   Esbern Friis-Hansen and Bhuwon Sthapit

6. Participatory approaches linking farmers' access to genebanks: Ethiopia 56
   Melaku Worede, Awegechew Teshome and Tesfaye Tesemma

7. Linking the national genebank of Vietnam and farmers 62
   Nguyen Ngoc De

8. Toward establishing links between farmers and the ICRISAT genebank 69
   Paula Bramel-Cox

9. Re-introducing crop genetic diversity in post-war Somalia 75
   Esbern Friis-Hansen, Dan Kiambi, Luigi Guarino and James Chweya

Section III. Local plant genetic resources management and participatory crop improvement

10. Overview 84
    Bhuwon Sthapit and Esbern Friis-Hansen

11. Experiences in participatory approaches to crop improvement in Nepal 90
    Anil Subedi, Krishna Joshi, Pratap Shrestha and Bhuwon Sthapit

12. Crop improvement at community level in Vietnam..103
    Nguyen Ngoc De

13. Mass selection: a low-cost, widely applicable method for local crop improvement
    in Nepal and Mexico..111
    Bhuwon Sthapit, Pratap Shrestha, Madhu Subedi and Fernando Castillo-Gonzales

14. Understanding farmers' knowledge systems and decision-making: participatory
    techniques for rapid biodiversity assessment and intensive data plots in Nepal...............117
    Ram Rana, Pratap Shrestha, Dipak Rijal, Anil Subedi and Bhuwon Sthapit

15. Participatory approaches to a study of plant genetic resources management
    in Tanzania..127
    Esbern Friis-Hansen

16. The role of gender in the conservation, location and management of genetic diversity
    in potatoes, tarwi and maize in Pocoata, Bolivia..131
    Lucio Iriarte, Litza Lizarte, Javier Franco, David Fernandez and Pablo Eyzaguirre

Section IV. Participatory approaches for establishing community seed banks and improving local seed systems

17. Overview..140
    Esbern Friis-Hansen and Bhuwon Sthapit

18. Community seed banks and seed exchange in Ethiopia: a farmer-led approach..............142
    Regassa Feyissa

19. Seed conservation and management: participatory approaches of *Nayakrishi*
    Seed Network in Bangladesh...149
    Farhad Mazhar

20. Local seed systems and PROINPA's genebank: working to improve seed quality
    of traditional Andean potatoes in Bolivia and Peru...154
    Victor Iriarte, Franz Terrazas Gino Aguirre and Graham Thiele

Section V. Increasing public and policy awareness of conservation and use of plant genetic resources

21. Overview..164
    Bhuwon Sthapit and Esbern Friis-Hansen

22. Adding value to landraces: community-based approaches for *in situ*
    conservation of plant genetic resources in Nepal..166
    Dipak Rijal, Ram Rana, Anil Subedi and Bhuwon Sthapit

23. Participatory ethnobotanical research for biodiversity conservation:
    experiences from Northern Nagaland, India..173
    Archana Godbole

24. Linking to community development: using participatory approaches to *in situ*
    conservation..181
    P.V. Satheesh

25. Identifying and analyzing policy issues in plant genetic resources management:
    experiences using participatory approaches in Nepal...188
    Devendra Gauchan, Anil Subedi and Pratap Shrestha

Acronyms..194

Glossary..196

Bibliography...204

Preface

IPGRI's revised strategy *Diversity for Development* (1999) recognizes the central role that farming communities, local cultures and institutions play in the conservation and use of plant genetic resources. Recognition of this fact was consolidated in various global fora from the Convention on Biological Diversity to the Global Plan of Action for Plant Genetic Resources which was approved by 155 countries in Leipzig Germany in 1996.

Translating this global policy on community participation into equitable practices in the fields, forests and villages where men and women farmers maintain and manage plant genetic resources (PGR) has led to the development and adaptation of a new set of tools for PGR conservation and use. Over the past five years, a wide range of partners in the informal sector, including both NGOs and civil society institutions, have joined with PGR scientists in conserving and enhancing the value of plant genetic resources in farming systems. This has had an important impact by focusing attention on a wider range of uses and diversity in crop and forest genetic resources. At the same time it has demonstrated the potential contribution that the deployment of plant genetic resources can make to the development and well-being of the rural poor, particularly those living in less favoured environments.

IPGRI considers this compilation of experiences, approaches and tools from around the world to be a significant step for consolidating the participation of rural communities in the conservation of agrobiodiversity. We place great value on the people-centred approach that holds plant genetic resources to be the essential biological assets of the rural poor. Through the various types of partnerships described in this book, there is a clear role for PGR programmes to contribute by adding value to those biological assets.

We look forward to continuing the research and experience with participatory approaches and to the consolidation of broad-based partnerships for the conservation and use of plant genetic resources. In the production of this work, many organizations and individuals throughout the world have contributed. We are particularly pleased to have worked with Denmark's Centre for Development Research.

Geoffrey Hawtin
Director General
IPGRI

Rome, June 2000

Acknowledgements

Our first acknowledgement is to the farmers who shared their time, knowledge and resources with the authors. This book was written over a two-year period, and we are grateful to those who helped in its preparation. The editors would especially like to thank the contributors who gave so much time and shared their valuable experiences in the making of this book. The production of this volume was supported by IPGRI and the Centre for Development Research (CDR) in Copenhagen, Denmark and we thank both institutions for their commitment to participatory approaches to agrobiodiversity.

We acknowledge the encouragement and advice of Pablo Eyzaguirre of IPGRI who saw the need to consolidate the technical experiences and methods of informal sectors and scientists using participatory approaches to PGR conservation and use. We would also like to extend our thanks for the advice and suggestions made by Ken Riley, Ramanatha Rao and Devra Jarvis of IPGRI, and Jannick Boesen of the "Nature, People and Society research theme" at CDR. Maria Fernandez, Louise Sperling and Jacqueline Ashby of the CGIAR Systemwide Program on Participatory Research and Gender Analysis (SPRGA) provided useful comments and a platform for developing this work. The administrative and technical support of Pratap Shrestha (LI-BIRD, Nepal), Amanda King (IPGRI Intern), Sajal Sthapit (Nepal), Jesper Linell and Anni Hammerlund (CDR) during the preparation of the manuscript is highly appreciated. We owe a great debt to Linda Sears, IPGRI's editor, whose editorial skills have shaped this volume in every way and to Patrizia Tazza for the cover art. In the course of this work we have built ties of friendship and trust in working together which made this a rewarding enterprise. For this we give profound thanks.

Esbern Friis-Hansen
Bhuwon Sthapit

Contributors

Aguirre, Gino
Head of Genetic Resources
Fundación para la Promoción e Investigación de Productos Andinos (PROINPA)
Casilla Postal 4285
Cochabamba, **BOLIVIA**
Email: PROINPA@papa.bo

Almekinders, Conny
CPRO/Centre for Genetic Resources, The Netherlands
PO Box 16
6700 AA Wageningen, **THE NETHERLANDS**
Tel: +31-317-4770076/477001
Fax:+31-317-418094
Email: c.j.m.almekinders@cpro.wag-ur.nl

Bramel-Cox, Paula
Principal Scientist
International Crops Research Institute for the Semi-Arid Tropics (ICRISAT)
Patancheru 502 324
Andhra Pradesh, **INDIA**
Tel: +91-40-3296161
Fax: +91-40-241239
Email: p.bramel-cox@cgiar.org

Castillo-González, Fernando
Professor, Mejoramiento Genetico de Maiz
Programa de Genetica
Colegio de Postgraduados
Instituto de Recursos Geneticos y Productividad
Km 36.5 Carretera Mexico-Texcoco
Montecillo, CP 56230, **MEXICO**
Tel: +91 595 10230
Fax: +91 595 11544
Email: fcastill@colpos.colpos.mx

Chewya, A. James †
Department of Crop Science
University of Nairobi/IPGRI
Nairobi, Kenya

De, Nguyen Ngoc
Head, Rice Research Department
Mekong Delta Farming Systems Research and Development Institute (MDFSRDI)
Can Tho University
Can Tho, **VIETNAM**
Tel: +84 71-831251/830040
Fax: +84 71-831270
Email: ltduong@ctu.edu.vn, rrd@ctu.edu.vn

De Boef, Walter
Agro-biodiversity & Development
Agriculture & Enterprise Development
Royal Tropical Institute (KIT)
Mauritskade 63
PO Box 95001
1090 Ha Amsterdam, **THE NETHERLANDS**
Telephone :+31 20 568 8483
Telefax: +31 20 568 8498
Email: w.d.boef@kit.nl
Web site: www.kit.nl

Eyzaguirre, Pablo
International Plant Genetic Resources Institute (IPGRI)
Via delle Sette Chiese 142
00145 Rome, **ITALY**
Tel: +39 06518921
Fax: +39 065750309
Email: p.eyzaguirre@cgiar.org

Fernandez, Maria
Senior Scientist
Systemwide Program for Participatory
Research and Gender Analysis
International Centre for Tropical Agriculture
(CIAT)
Tripoli 350/401
Lima 18, **PERU**
Email: m.fernandez@cgiar.org

Feyissa, Regassa
Institute of Biodiversity Conservation and
Research (IBCR)
PO Box 30726
Addis Ababa, **ETHIOPIA**
Tel: +251 1 615607/612244
Fax: +251 1 613722

Friis-Hansen, Esbern
Senior Researcher
Centre for Development Research
Gl Kongevej 5, Copenhagen 1610V, **DENMARK**
Tel. +45 3385 4600 Direct line +45 3385 4646
Fax. +45 3325 8110
Email: efh@cdr.dk
Web site: www.cdr.dk

Gauchan, Devendra
Senior Scientist (Agricultural Economist)
Outreach Research Division
Nepal Agricultural Research Council
PO Box 5459
Lalitpur, **NEPAL**
Tel: +977 1 540817
Fax: +977 1 521197
Email: dgauchan@hotmail.com,
 cpdd@mos.com.np

Godbole, Archana
Project Co ordinator
Applied Environmental Research Foundation
Ganga Tara Apts. 917/7, Ganeshwadi D.G.
Near British Council Library
Pune- 411 004, **INDIA**
Tel. +91 020 350567
Email: aerf@pn2.vsnl.net.in

Guarino, Luigi
Scientist, Genetic Diversity
International Plant Genetic Resources
Institute (IPGRI)-Americas
C/O CIAT Apartado Aereo 6713
Cali, **COLOMBIA**
Tel: +57 2 445-0048/445-0049 IPGRI
Fax: +57 2 445-0096 IPGRI
Email: l.guarino@csi.cgiar.org

Itriarte, Victor
Fundación para la Promoción e
Investigación de Productos Andinos
(PROINPA)
Casilla Postal 4285
Cochabamba, **BOLIVIA**
Email: PROINPA@papa.bo

Joshi, Krishna Dev
Programme Officer (Plant breeder)
Local Initiatives for Biodiversity, Research
and Development (LI-BIRD)
PO Box 24, Narayangadh, Chitwan, **NEPAL**
Tel/fax: +977 56 21029
Email: libird@mos.com.np

Kiambi, Dan K.
Scientist, GRST
IPGRI Regional Office for Sub-Saharan
Africa
c/o ICRAF
PO Box 30677
Nairobi, **KENYA**
Tel: (254-2) 524509
Fax: (254-2) 524501
Email: d.kiambi@cgiar.org

King, Amanda
Ethnobotany intern, IPGRI, Rome
Current address:
3057 Hillegass Ave.
Berkeley, CA 94705, **USA**
Tel: +1 (510) 666-8423
Email: aking@nature.berkeley.edu

Technical contributors

Mazhar, Farhad
UBINIG
5/3 Barabo Mahanpur, Ring Road
Shaymoli, Dhaka-1207, **BANGLADESH**
Tel: +880 2 811465/3296209
Fax: +880 2 813065
Email: ubinig@citechco.net

Rana, Ram Bahadur
Programme Officer (Socio-economist)
Local Initiatives for Biodiversity, Research
and Development (LI-BIRD)
PO Box 324, Banstolathar
Pokhara, **NEPAL**
Tel/Fax: +977 61 26834
Email: rblibird@mos.com.np,
libird@mos.com.np

Rijal, Dipak Kumar
Programme Officer
Local Initiatives for Biodiversity, Research
and Development (LI-BIRD)
PO Box 324, Banstolathar
Pokhara, **NEPAL**
Tel/Fax: 977 61 26834
Email: drlibird@mos.com.np,
libird@mos.com.np

Satheesh, P.V.
Director
Deccan Development Society
A-6, Meera Apartments, Basheerbagh
Hyderabad – 500 029, **INDIA**
Tel: +91 40 3222867/3222260
Fax: +91 40 3222260
Email: ddshyd@hd1.vsnl.net.in

Shrestha, Pratap Kumar
Programme Officer (Social scientist)
Local Initiatives for Biodiversity, Research
and Development (LI-BIRD)
PO Box 324, Banstolathar
Pokhara, **NEPAL**
Tel/Fax: +977 61 26834
Email: pslibird@mos.com.np,
libird@mos.com.np

Sthapit, Bhuwon Ratna
Scientist
IPGRI-APO
PO Box 236 UPM Post Office
Serdang 43400, Selangor Darul Ehsan,
MALAYSIA
Tel: +603 942 3891
Fax: +603 948 7655
Email: b.sthapit@cgiar.org,
sthapit@mos.com.np

Subedi, Anil
Executive Director
Local Initiatives for Biodiversity, Research
and Development (LI-BIRD)
PO Box 324, Banstolathar
Pokhara, **NEPAL**
Tel/Fax: +977 61 26834
Email: aslibird@mos.com.np,
libird@mos.com.np

Subedi, Madhu
Programme Officer (Plant breeder)
Local Initiatives for Biodiversity, Research
and Development (LI-BIRD)
PO Box 324, Banstolathar
Pokhara, **NEPAL**
Tel/Fax: +977 61 26834
Email: mslibird@mos.com.np,
libird@mos.com.np

Terrazas, F.
Fundación para la Promoción e
Investigación de Productos Andinos
(PROINPA)
Casilla Postal 4285
Cochabamba, **BOLIVIA**
Email: fterraza@proinpa.org

Teshome, Awegechew
Associate Scientist
IPGRI, Rome, **ITALY**
Current address:
30 Marie Curie
Biology Department, University of Ottawa
Ottawa, ON, **CANADA** K1N 6N5
Email: ateshome@science.uottawa.ca

Tesemma, Tesfaye
International Seeds of Survival Programme
(SoS)
PO Box 62857
Addis Ababa, **ETHIOPIA**
Tel: +251 1 530925
Fax: +251 1 -530925/510672

Thiele, Graham
Proyecto Papa Andina CIP-COSUDE
Casilla Postal 4285, Cochabamba, **BOLIVIA**
Tel: +591 360800/360801
Fax: +591 360802
Email: g.thiele@cgiar.org

Worede, Melaku
Scientific Advisor
International Seeds of Survival Programme
(SoS)
PO Box 62857
Addis Ababa, **ETHIOPIA**
Tel: +251 1 530925
Fax: +251 1 530925/510672

Introduction

Esbern Friis-Hansen and Bhuwon Sthapit

Since the 1992 UNCED conference on environment and development, there have been increasing political calls for broad-based and holistic approaches to the conservation and use of plant genetic resources. The successful implementation of such conservation approaches involves working more directly with farmers. This is reflected in the Global Plan of Action that was approved at the FAO International Technical Conference on Plant Genetic Resources for Food and Agriculture in Leipzig 1996.

> "Strengthening farmers' community level management of plant genetic resources is essential to the success of *in situ* conservation and development, and to facilitate the sharing of benefits derived from the utilization of these resources. Farmers and their communities play a critical role in the conservation and improvement of plant genetic resources for food and agriculture. Enhancing their capacity would help promote food security, particularly among the many rural people who live in agriculturally-marginal regions." (FAO Global Plan of Action 1996:15).

Community participation in the conservation and use of plant genetic resources is not new. It is an integral part of the management of natural resources by many agrarian communities. What is new is the recognition by formal-sector plant genetic resources institutions that there are areas, particularly involving *in situ* conservation, species with multiple uses, and neglected and underutilized crops where community-based management of plant genetic resources is an effective way to conserve agrodiversity. While direct compensation to farmers is not intended, it is important that the global investment in farmer's welfare, through local initiatives of management of local genetic resources, be seen as indirect compensation in recognition of their role in on-farm crop conservation. This kind of indirect compensation may reach more farmers and thus be more equitable than a system of payment to a few farmers.

While a number of NGOs, international centres and national agricultural research systems, including plant genetic resources programmes, have experimented with participatory approaches in their projects relating to plant genetic resources conservation and use, much work on developing participatory methodologies which are adapted to specific activities still needs to be done and documented. The initial results of participatory approaches to plant genetic resources conservation and use have been compiled and published in the proceedings of conferences held during the past three years (see, e.g. Eyzaguirre and Iwanaga 1996; Sperling and Loevinsohn 1996; UPWARD 1996; CIAT 1997; Veldhuizen *et al.* 1997). Recently, the CGIAR Programme on Participatory Research and Gender Analysis (PRGA) has also published case studies of participatory approaches on plant breeding (McGuire *et al.* 1999).

When one reads through the more than one hundred case studies and articles included in these five recent publications and additional studies and project documents published separately, one finds enormous variety in the approach to participation and its practice. Participatory approaches range from collection of anecdotal information during occasional visits to farmers, to rigorous work which fundamentally challenges the conventions of western empiricism which still underpin most applied agricultural research, and which have demonstrated the potential to revolutionize the way in which public-sector agricultural research serves resource-poor farmers in difficult environments (CIAT 1997).

For most national plant genetic resources programmes, the involvement of farmers as partners in a dialogue would constitute a dramatic change to the existing approaches to plant genetic resources conservation and use. The conventional *ex situ* conservation approach is based on an extraction of germplasm from farmers for storage in *ex situ* genebanks, and linked to agricultural modernization programmes based on conventional top-down transfer of technology from research through extension to farmers. Today, such top-down approaches are often complemented by the use of generalized information about the farming system (farming systems research and inclusion of socioeconomic information data in germplasm collection missions). The conventional approach nevertheless still considers farmers as passive players and all research and innovations are carried out and implemented by researchers.

In contrast to the conventional methods, the new approaches to conservation and use of plant genetic resources are based on a high level of participation of farmers and their organizations at the local level. The purpose of the exercise is not a mere "physical" participation but the participatory approach aims to take comparative advantages of both scientific and indigenous knowledge systems. There are still, however, a number of obstacles to the proliferation of participatory programmes. There are many ethical, policy and institutional factors which can lead to situations where participation by communities is discouraged, or where local community contributions are ignored because of the formal research community's different values and approaches to conservation and use. Participatory community-based methods can thus easily be experienced by conventional scientists as haphazard ventures, over which they quickly lose control.

Participatory approaches to plant genetic resources conservation and use are thus still in the initial stages of development and it is too early to draw conclusions or implement guidelines. This technical bulletin will present a number of case studies that will enhance the understanding of the multitude of participatory approaches to conservation and use of plant genetic resources which are emerging.

Organization of the bulletin

The bulletin is divided into five sections. The first section serves as an introduction to the crosscutting issues related to integrating gender analysis and institutional context. It provides a brief overview of the concepts, processes and tools associated with participatory approaches, followed by five sections illustrating different forms of participatory approaches within the areas of plant genetic resources management through selected case studies. Each section starts with an overview of issues involved, written by the editors, followed by several case studies. Section II documents four case studies that enhance access of germplasm and information to farming community from the genebank. Section III introduces various ways and methods of local plant genetic resources management at community level. New participatory approaches to assess biodiversity and understand reasons of conservation and use have been outlined. There are a number of efforts under way to encourage a wide scope for farmer participation in formal plant breeding. The section reviews recent experiences with farmer participation in variety selection and plant breeding as well as managing diversity in farmers' fields. Section IV examines a few participatory approaches to community seed bank management and seed exchange and its role in on-farm conservation. Finally, Section V is based upon participatory methods of increasing public and policy awareness of conservation and use of plant genetic resources. The section describes the challenges to crop genetic diversity, presents some of the value-addition strategies that are being implemented to reverse the erosion of the diversity, and concludes with some policy implications.

References

CIAT. 1997. New Frontiers in Participatory Research and Gender Analysis. CGIAR Systemwide Program on Participatory Research and Gender Analysis for Technology Development and Institutional Innovation. CIAT, Cali, Colombia.

Eyzaguirre, P. and M. Iwanaga, eds. 1996. Participatory Plant Breeding. Proceedings of a workshop on participatory plant breeding, 26-29 July 1995, Wageningen, The Netherlands. International Plant Genetic Resources Institute, Rome.

McGuire, S., G. Manicad and L. Sperling. 1999. Technical and Institutional Issues in Participatory Plant Breeding - Do from a Perspective of Farmer Plant Breeding. A Global Analysis of Issues and Current Experience. CGIAR Systemwide Program on Participatory Research and Gender Analysis for Technology Development and Institutional Innovation. CIAT, Cali.

Sperling, L. and M. Loevinsohn, eds. 1996. Using Diversity: Enhancing and Maintaining Genetic Resources On-Farm. Proceedings of a workshop held on 19-21 June 1995, New Delhi, India. IDRC, New Delhi.

UPWARD. 1996. Into Action Research, Partnerships in Asian Rootcrop Research & Development. UPWARDS, Manila.

Veldhuizen, Laurens van, Ann Waters-Bayer, Ricardo Ramirez, Debra A. Johnson and John Thompson (eds.) 1997. Farmers' research in practice. ILEIA Readings in Sustainable Agriculture. Intermediate Technology Publications, London.

Section I

Crosscutting issues

1. Concepts and rationale of participatory approaches to conservation and use of plant genetic resources

Esbern Friis-Hansen and Bhuwon Sthapit

Introduction to the concept of participation

Participatory approaches, it is argued, ensure greater efficiency and effectiveness of investments and contribute to a process of empowerment of the participants. A wealth of participatory techniques has emerged over the past 15 years in relation to research and development programmes. Among important organizations which have contributed to the development of such participatory methodologies are: ODI (London), IIED (London), IDS (Sussex), Clark University Program for International Development and Social Change (Melissa Leach, Worchester, US), WRI (Washington DC) and MYRADA (Bangalore).

Many of the participatory techniques developed at these institutions, and elsewhere, can be adapted and applied successfully to activities of plant genetic resources conservation and use. While much of this work has yet to be done, case studies to date provide many good examples of the practical use of participatory techniques and tools. There is no single correct method of management for successful participatory research. Rather, the choice of participatory approach will depend on the point of departure and the objective of the research and development.

There is considerable confusion concerning the definition of participation and the practical implementation of participatory development programmes (Oakley 1991). The differing views reflect different research traditions and assumptions about the relationship between participation and development. During the 1950s and 1960s, political participation in development was largely considered to be an obstacle to efficient economic growth. In the 1970s, mainstream views changed and participation was increasingly regarded as a necessary condition for economic growth and a means to ensure a minimum measure of equity. In development programmes, participation became associated with the Basic Needs Strategy.

For agricultural research, participation became associated with Farming Systems Research in the 1980s. Since the late 1980s, participation has become an integrated element of sustainable development strategy. The critical importance of participation to sustainable development has become widely accepted within the UN and among international donor organizations. In the agricultural arena adaptive and farming systems research on agricultural research stations began to include user perspective analysis in the mid-1980s. Substantial work has since been done to refine participatory agricultural research methodologies, not least within the CGIAR system, e.g. On-farm Research Methodology (CIMMYT), Women and Rice Farming Network (IRRI), Farmer Participatory Research (CIAT), On-farm Research (IITA), and Farmer-Back-To-Farmer Approach (CIP).

The literature distinguishes between two different rationales for research and development organizations to engage in participatory approaches: improved cost-effectiveness and empowerment.

In the development debate, the motivation to engage in participatory approaches for many organizations (who are concerned about cost-effectiveness) is that the projects are thought more likely to be successful if the beneficiaries are actively involved. Farmers and others who participate actively in a project are likely to be more committed, facilitating acceptance of policies and technologies promoted by outside agents. Participation of beneficiaries is also thought to enhance the ability of projects to exploit the potentials of indigenous technical knowledge. In the case of research, increased participation of farmers is thought to increase the functional efficiency of formal-sector agricultural research activities, in particular in reaching poor farmers in marginal areas. The rationale for increased participation in this context is that science can make its most effective contribution to research and development for the poor when it takes account of and utilizes farmers' indigenous knowledge based systems.

A participatory approach to improvement of farmers' plant genetic resources management offers the potential to reach a large number of farmers and make local crop genetic diversity an integral component of agricultural development for farmers in less-favoured regions. Achieving this goal necessarily depends on farmers themselves and requires building upon and making use of farmers' ongoing efforts to improve their crops through mass selection and other breeding efforts. It also requires recognizing the central role that rural women play in agricultural production in most developing countries. Programmes which provide farmers greater access to appropriate genetic resources and training could assist farmers in improving various characteristics of their planting materials (such as disease or pest resistance and drought tolerance) and in increasing food production. A number of governments, research institutions and NGOs are now engaged in projects researching and promoting on-farm management and improvement of plant genetic resources.

Other organizations view participation as a means and an end in itself. Participation is thought to empower poor people by enhancing local management capacity, increasing confidence in their own abilities as individuals as well as a community. NGOs and CBOs, in particular, view participatory approaches to plant genetic resources conservation and use as a means of empowering poor farmers by strengthening their indigenous capacity to manage local PRA techniques and to formulate effective demands for external science-based assistance. Participatory cooperation may empower individual farmers, as well as their formal and informal institutions for plant genetic resources management, and may enhance individual farmer's ability to develop their local farming systems.

Empowerment is often used in an unclear way as an argument for a participatory approach. In many participatory development projects, empowerment is used in an apolitical way, in which the individuals are empowered to choose opportunities offered by the project. The mechanisms of empowerment in most participatory development projects are either very simple, such as involvement in the market economy, or conveniently fuzzy. There is a need for a better understanding of the social structures and processes of the societies in which participatory approaches are applied (Long and Long 1992).

As participation has become an integral part of mainstream development activities, a wealth of different forms and levels of participation has emerged. There have been several attempts to classify different forms of participation, and we will in the following refer to two of these.

White (1996) classified participation into four categories (Table 1.1). The lowest level is nominal participation, where the farmer simply lends/rents land and labour to the researchers. The second level is consultative participation, where farmers' opinions and knowledge are explored. The third level is action-oriented participation, where farmers are directly involved with implementing part of the research activities. The fourth level is decision-making/design participation, in which farmers take part in deciding the objectives of the research, plan the design of the experiments and implement them. The level of participation chosen will have significant consequences for the research process and methodology. White's classification attributes different interests to external agents and beneficiaries for each level of participation. She argued that the external development agents and the beneficiaries only share the same interest (empowerment) in the case of transformative participation.

However, it is not only important to classify different forms of participation, but equally important to differentiate between who participates and how. Uphoff *et al.*'s (1979) classification of participation (Table 1.2) incorporates such considerations. The kind of participation is based on the project cycle, but the project phases are not necessarily sequential. The classification of who participates distinguishes four different actors. Finally the table distinguishes between different mechanisms by which participation takes place.

Table 1.1. Interests in participation

Form of participation	Interest of external agent	Interest of beneficiaries	Function
Nominal	Instrumental	Representative	Transformative
Legitimization	Efficiency	Sustainability	Empowerment
Inclusive	Cost	Leverage	Empowerment
Display	Means	Voice	Means/end

Source: After White (1996).

Table 1.2. Dimensions of rural development participation

Kinds of participation	Participation in decision-making
	Participation in implementation
	Participation in benefits
	Participation in evaluation
Who participates ?	Local residents
	Local leaders
	Government personnel
	Foreign personnel
How is participation occurring ?	Basis of participation
	Form of participation
	Extent of participation
	Effect of participation

Source: Uphoff *et al.* (1979).

Review of experiences using participatory approaches

Validation of the benefits of participatory approaches

Only limited evidence of long-term effectiveness in terms of improvement of living conditions of participatory development interventions exists and the evidence available is concerned with projects on a small scale only. Evidence of empowerment and sustainability, based on proven outcomes, is even more scattered. In the absence of evidence from monitoring and evaluations, engagement in participatory approaches by outside development organizations becomes an act of faith that involvement of people will uncover the real picture and that technologies which are appropriate will be chosen or developed.

The approach to establishing a dialogue between scientific and local experience-based knowledge is one of the first, and perhaps the most important, choices facing any participatory research programme on plant genetic resources. Participatory approaches require the combining of natural and social sciences with local experience-based knowledge. There are two possible points of departure for establishing the necessary dialogue. Scientists can take local farming systems and farmers' indigenous experimental practices as their point of departure and build on to and support experience-based knowledge with scientific-based knowledge and methods. Alternatively, scientists can take their own science-based experiments as a point of departure and then involve farmers. Departing from a strictly science-base starting point creates a range of methodological problems for scientists (e.g. how to assess farmer-defined and managed experiences which are often too complex to enable statistical analysis).

Most participatory research and development projects relating to plant genetic resources have chosen the second of the starting points outlined above. There are at least two possible reasons why this is so. First, most participatory research relating to plant genetic resources is carried out by foresighted individuals working at otherwise conventional research stations. Objectives therefore commonly focus on the development and adoption of a product (e.g. involving farmers in screening/ testing pre-released varieties to ensure high adoption rates) rather than the process (e.g. strengthening a farming communities' indigenous capacity for managing its own plant genetic resources). Second, knowledge and understanding of farmers' indigenous plant genetic resources management practices is still limited and the concept of science-aided indigenous agricultural research is alien to most agricultural scientists.

Deciding on issues suitable for participatory cooperation

When identifying issues suitable for participatory cooperation, it is important to analyze the areas of competence, expertise and interest of the communities concerned with the management of plant genetic resources. Farmers' decisions on selection, treatment, cultivation, processing and end-use of varieties are crop specific. Management of diversity may vary among different social, gender, ethnic and age groups of farmers within the community. Understanding farmers decision-making with regards to managing intra-specific diversity of crops requires an analysis of the wider context in which farming takes place including, among other things, the dynamics of farming systems and policies on agricultural modernization.

Stakeholder analysis

Participation is time-consuming and any research programme has to select a limited number of representatives among stakeholders to participate. Who and how many are to participate, how they are differentiated (in social, gender and cultural terms), at which state in the research process and with what weight, are important questions which implicitly or explicitly are addressed during any participatory research programme and which have consequences for the level of participation.

Participatory research is often multidisciplinary, involving plant genetic resources specialists and social scientists. An essential analysis, often undertaken by social scientists carrying out participatory research, is to identify the stakeholders. Stakeholders are not only those who are directly involved in carrying out the research but also those who are the intended users of the research results and others who might help shape the research agenda (CIAT 1997: Introduction).

There is a general consensus that the best way of working with farmers is to take a group approach rather than to work with individual farmers. A group approach increases the efficiency of the outside researchers' work and strengthens the local capacity to innovate. A participatory research project's choice of collaborating indigenous institutions has to be based on a thorough survey of the existence and character of local institutions which have an influence on agriculture and which could facilitate participatory cooperation. A village often contains several organizations with overlapping goals and memberships which may represent competing interests. It is thus crucial that researchers do not simply chose to cooperate with the established village leadership, which may not be representative of all social and cultural groups and may not adequately represent the interests of women.

Organizing farmers to commit themselves to continue specific systems of managing species diversity is not an easy task and no standard organizational model exists. A number of principles may be drawn from reviewing the experience gained from community development projects and *in situ* projects in particular. As *in situ* conservation projects depend on farmers' voluntary participation based on a combination of incentives and self-interest, the leaders of farmer organizations at the local level must be legitimate and must include those individuals who in practice take crop management decisions. They are not necessarily the male head of household, local politicians, or other persons in the community who are better off and/or have power and status in the community. Organizing farmers for *in situ* conservation may prove to be particularly difficult in communities where intra-species diversity is cultivated by the poor, while the relatively better off, who often constitute the local political elite, cultivate modern varieties.

> **Participatory approaches to natural resource management**
>
> Participatory approaches to natural resource management projects, including water, forestry and rangeland management, became common among government institutions, donor organizations and NGOs during the 1990s. Typical development narratives of participatory natural resource management projects regard most communities as having once been able to regulate natural resource use and technology in a way that society and environment remained in equilibrium. For various reasons, including population growth, breakdown of traditional authority, commercialization, agricultural modernization, or social change caused by implementation of inappropriate state policies, such equilibrium is no longer present. Participatory natural resource management (NRM) projects aim to re-establish equilibrium between the community and the environment by supporting the community in its efforts to organize itself and use its knowledge and skill in a beneficial manner.
>
> Participatory approaches to natural resource management have often failed, both from the point of view of the implementing agencies (e.g. formal government institutions, donor organizations or NGOs) and from the involved communities' point of view (Pimbert and Pretty 1995). Critics suggest that there are key problems related to the participatory projects underlying assumptions (implicit or explicit) that local communities are relatively homogeneous and that local organizations, e.g. village environmental committees, are legitimate representatives of all people in the community. Local communities are often socially diverse. Local natural resource management practices are influenced by a number of social factors, including wealth, ethnicity, gender, age and settlement history. Pastoralists have different views on land use than agriculturalists and relatively well-off farmers engage in different natural resource management practices than poor farmers. How men and women are involved in natural resource management is closely related to the gender division of labour.
>
> Conflicts over natural resource management are widespread, reflecting diverse values and resource priorities among different actors. While natural resource management conflicts have long been discussed within social sciences, such conflicts have largely been ignored by most government and donor projects with participatory approaches to natural resource management.
>
> To summarize, formal organizations at the local level (e.g. village government, representatives of central ministries) are often problematic as the focal point for participatory natural resource management projects because: (1) such organizations often do not represent the interests of all members of a given community and often reproduce relations of unequal power and authority, and (2) the power and influence of such organizations over actual natural resource management practices is limited.

Identification and empowerment of local institutions for the *in situ* conservation of landraces links into a wider debate on the role of local institutions in resource management. Issues include land tenure, sharecropping arrangements and common grazing rights, among others. The concept of local institutions as used here is to be understood broadly as the content and dynamics of norms and traditions within a community for managing their resources (not to be confused with formal-sector organizations dealing with natural resource management).

Gender differentiation

The gender division of labour is highly relevant for selecting which farmers should participate in plant genetic resources activities. It is impossible to generalize about gender divisions of labour, as these vary greatly and are often highly local or ethnic-specific. However, it is generally recognized that women often play a key role in domesticating wild species, selecting, processing, storing and exchanging seed. The gender division of labour is often crop-specific and women commonly dominate the management of food crops which are cultivated primarily for household food consumption. The gender issue will be discussed in Chapter 3.

Cooperation with community institutions

Participatory approaches are often influenced by new institutional theories, which argue that institutions help to formalize mutual expectations of cooperative actions, allowing penalties for non-compliance and thereby limiting the extent of free riding to an acceptable level. The World Bank has become a leading exponent for such 'new insitutionalism'. While recognizing the importance of informal institutions, most international donor organizations have focused on building formal institutions (Uphoff 1992) when implementing participatory projects.

There is a need to understand local norms and processes of decision-making and how these change and are negotiated and in particular how some people may influence outcomes with direct participation. If not, participatory institutions which are formalized through external interventions risk becoming empty shells, with important decisions and collective actions taking place elsewhere.

The strength of local institutions depends on the time and resources that people are willing to invest in them. People's support of local institutions is closely related to whether they are viewed as legitimate and represent their interests. Local institutions do not necessarily take the form of an organization with an aim and agenda but may simply consist of a group of people regulating natural resources over time. Different social groups may simultaneously support different rules and institutions at a given location. The actual entitlement of an individual is affected by the interaction between such competing notions of legitimacy. While formal institutions, such as state regulations, are enforced by representatives of formal organizations, local institutions exist through the agreement and support of the local actors involved.

References

CIAT. 1997. New Frontiers in Participatory Research and Gender Analysis. CGIAR Systemwide Program on Participatory Research and Gender Analysis for Technology Development and Institutional Innovation. CIAT, Cali, Colombia.

Long, N. and A. Long (eds.). 1992. Battlefields of knowledge: the interlocking of theory and practice in social research and development. Routledge, London.

Oakley, P. 1991. Projects with People: The Practice of Participation in Rural Development. ILO, Geneva.

Pimbert, M.P. and J.N. Pretty. 1995. Parks, people and professionals: putting "participation" into protected area management. UNRISD Discussion Paper no. 57. UNRISD, Geneva.

Uphoff, N.T., J.M. Cohen and A.A. Goldsmith. 1979. Feasibility and application of rural development participation: a state-of-the-art-paper. Cornell University, New York.

Uphoff, N.T. 1992. Local institutions and participation for sustainable development. IIED Gatekeeper Series no. 31. IIED, London.

White, Sara C. 1996. Depoliticising development: The uses and abuses of participation. Development in Practice, Vol. 6, No. 1. Oxfam, UK and Ireland.

2. Institutional perspectives on participatory approaches to use and conservation of agrobiodiversity

Conny Almekinders and Walter de Boef

Introduction

This presentation discusses the institutional functions and relations in the use and conservation of Plant Genetic Resources (PGR). The system in which PGR is managed can be seen as consisting of two subsystems: a local system in which farmers are the principal actors, and an institutional system, in which genebanks and institutions with breeding and seed programmes are the main actors. To support the agricultural development and sustainability of the local PGR system, effective support from the formal system is needed. Standard model approaches have not been very effective; other types of support need to be designed, implemented and monitored for increased impact. Complementarity of the farmer and institutional systems suggests that development of more, and better-integrated, linkages between these two systems can increase the effectiveness of both systems. An essential element of these linkages is the capacity of the institutional system to respond to farmers' needs for diversity. In the approaches to link with the local system, participation of and collaboration with farmers is essential.

The farmers' and institutional system

In plant genetic resources management a farmer's system and an institutional system[1] can be distinguished (de Boef *et al.* 1997). Historically, the farmer's system was the only one, emerging as an element of agriculture itself. In this system, farmers produce their own seeds on their farm, or obtain seed via exchange or purchase from other farmers, relatives or local middlemen. This farmer's system has elements of crop conservation, crop development and seed supply. These elements are well integrated, usually incorporated in the normal crop production, and all carried out on the individual farm or by the community as a whole (Fig. 2.1). Farmers' practices and selection criteria in the system of local seed reproduction define the varieties used and the shape of the geneflows within and between crop variety populations. It is therefore a local PGR system, managed in an integrated manner by farmers; it is still the prevailing system in the world. The system is generally characterized as a system which also harbours relatively large amounts of genetic diversity. For the farmers this diversity is functional in dealing with the variable environment and an unpredictable future (Clawson 1985; Prain and Hagmann 2000).

Parallel to this system an institutional system developed. This development started when genes were 'discovered' and the insights developed on how, with controlled crossing of selected parents, the combination of desired crop characteristics could be directed. Other, more complicated schemes than those based on mass selection, mutation and spontaneous outcrossing resulted in more effective crop development. Plant breeding, as the applied form of genetics, became a specialized activity, taking place outside the farmers' production system. Specialized non-farmers (plant breeders) carried out the crossing and selection activities off-farm. Genebanks were set up as institutions which specialized in the maintenance of large collections of base material for breeders. The objective of genebanks as repositories of genetic material threatened with loss emerged several decades later, when it was clear that breeders' success replaced large amounts of genetic diversity. Seed programmes are the institutions at the end of the formal PGR chain with the principal mandate to diffuse the improved materials coming out of the breeding programmes. Distribution targeted in the first place the farmers in the high-potential areas who used seed as an external input in agricultural production.

The limitations of the institutional system

The institutional system proved a successful formula for agricultural development in North America and NW Europe. In high-potential, more uniform areas in developing countries the formula was quite successful as well (SE Asia, irrigated rice), but it failed in more marginal areas (Almekinders *et al.* 1999). Introduction of new varieties resulted in partial or no adoption at all, seeds were often of similar or lower quality than the farmers' seed, and seed was in many cases not available at the time needed or was too expensive for the farmer. A range of factors contributed to a limited impact of the formal systems in the low-potential areas: limited accessibility (remote locations, bad roads, etc.); lack of knowledge and understanding of farmers' needs, preferences and production conditions; large and unpredictable environmental variation over time and space in the target areas (producing large Genotype x Environment effects), and the relatively small size of each target area with a specific combination of conditions and needs. The institutional system (with separate organizations or programmes responsible for conservation, breeding and seed supply) had developed as a system which runs parallel to the farmers' system with only two intentional points of contact: the genebank collecting missions and the distribution of improved seeds. These two points represent, respectively, a flow out of the local system and a flow into the local system (Fig. 2.1). The latter linkage is particularly weak owing to the mismatch between the farmers' needs on the one side, and the offered technology of the institutional system on the other. One aspect of the mismatch is that formal breeding and seed programmes normally produce and distribute seeds of only a few improved varieties. Farmers in the more variable low-input environments usually need more diversity: they plant more than one variety per crop, often including local varieties, and varieties which are genetically variable. Thus, the flows of genetic materials between the two systems are unbalanced: it is principally a flow out of the system with no substantial flow coming back.

A history of failing impact of the formal breeding and seed institutions has, however, a positive side as well: traditional genetic diversity has been only partially replaced by genetically more uniform improved materials, or not replaced at all. The impact varies from area to area, and from crop to crop (Byerlee 1994). The majority of the farmers in developing countries still grow traditional varieties, sometimes next to improved varieties, in other cases not. Traditional varieties represent a pool of genetic diversity that is adapted to local conditions through repeated reproduction and selection by farmers. One part of this adaptation is the crop genetic variation represented by the diversity of varieties that farmers grow. Another part of this adaptation is the genetic variation within the crop populations: this variation buffers crop yields against the unpredictable environmental conditions. Thus, many of the more marginal areas in the developing countries remain 'cradles' of genetic diversity. Many of the world's agricultural crops were domesticated in such environments and are therefore given prime importance by conservationists. Traditional varieties thus represent a valuable resource for farmers and humankind to deal with current and future environmental variation and uncertainty. However, in many situations they also lack acceptable yield capacity. They are often vulnerable to diseases that gained importance with changing agricultural conditions, and lack biotic stress tolerance or other specific valuable traits that are found in the genes of materials elsewhere in the world.

The arguments presented above explain why the attention of both the development and conservation-oriented PGR activities is now focusing on those lesser-developed, more marginal areas. Neither traditional nor improved varieties fully meet the farmers' needs. Farmers in these areas have, however, shaped and still maintain large amounts of basic genetic materials that are of importance for future agricultural development in the world. This characterization probably

---

[1] In this paper, 'farmer' and 'local' system are considered synonyms; similarly the terms 'formal' and 'institutional' system are considered synonyms.

accounts for the major part of the agricultural land in the developing countries. The farming systems in these areas are under pressure. A lagging development in combination with increased population pressure and increasingly unfavourable price ratios for food producers led in many cases to migration, unsustainable production practices, and reduced the capacity of the farmers to maintain genetic diversity.

Opportunities for the institutional PGR system

To properly address the farmers' need for diversity, the institutional PGR system has to re-orient its approach. To serve different target areas and develop a wider portfolio of materials, breeding programmes require differentiation in terms of (combination of) criteria and locations of selection. Seed production and diffusion schemes for such areas can only be effective when recognizing that farmers have different demands for seed than farmers in high-input areas where market production for the market dominates. Organization of such breeding and seed supply requires a more decentralized set-up than the current centralized standard model. Given the resources available, this can be achieved only by building on complementarity and collaboration with farmers and other organizations.

An analysis of strengths and weaknesses of the institutional and farmers' systems reveals many complementarities that offer opportunities for collaboration and farmer participation. First of all, farmers know their own production conditions and priorities. Farmers are often, but not always, very capable of adapting genetic material to local conditions through seed selection, of producing good-quality seed and diffusing materials via traditional exchange mechanisms. This does not mean that these farmers' practices cannot be improved. Rather, they are the practices that offer opportunities for further development and that are the basis for new initiatives. Farmers' systems usually have fewer opportunities for the introduction of exotic genes and genetic recombination, and are often performing poorly in the production of disease-free seed and in seed storage. In contrast, the formal system has much to offer in terms of access to new genes, techniques for genetic recombination, and knowledge and insights on seed technology aspects. Weaknesses of the formal system are the lack of knowledge and understanding of local conditions and farmers' preferences, and the limited possibility to address many needs and work in a range of conditions.

A logical conclusion follows from the above: the formal institutional system is likely to be more effective when focusing on the aspects in which they are strong, and should look for collaboration on the points in which they are weak and others are strong. Looking for farmers' participation in breeding and seed programmes is a logical and strategic step (Hardon and de Boef 1993). However, from strategy to practice is another step that is more difficult to make. Participation has many different forms and each situation has its particular conditions, crops, farmers and local organizations. Each situation is unique: the researchers have to analyze what is needed and what is a promising approach. The involved researchers have to design a locally specific, adapted approach for each activity. This includes analysis of involved stakeholders and their needs and interests, identification of collaborating farmers and organizations, definition of planning, monitoring and evaluation procedures. The cases in this publication are examples of such location-specific activities.

It is good to realize that these cases are indeed examples and not yet mainstream. For these activities to become 'mainstream' in the institutional organizations, approaches need to be formulated for upscaling the activities and approaches. These approaches need to give more space for diversity in development, i.e. space for genetic diversity, cultural diversity and for diversity in approaches. Based on the foregoing, a few points of consideration in relation to collaborative linkages do need attention. In an up-scaled situation, the involved researchers such as those presenting their experiences in this book, do need to look for collaborating organizations. Genebanks, for instance, have a limited number of staff members while a network of community

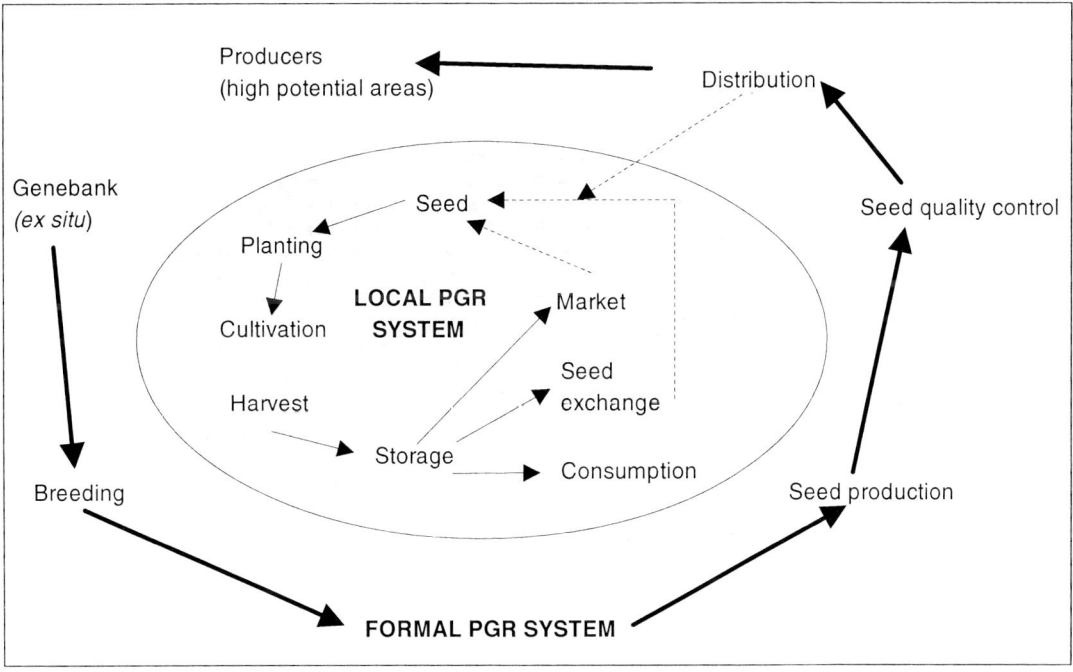

Fig. 2.1. Local and formal systems of management of plant genetic resources. Farmers are managing their plant genetic resources in an integrated and adaptive way. Crops are planted and harvested with a multiple purpose: to produce for household consumption and other on-farm uses, for marketing and to produce the seed, roots, tubers or stems for next-season planting. The exchange and adoption of seeds is another element that adds to the dynamics of this system. Seed is exchanged with relatives, friends or via merchants. Farmers may also use produce meant for consumption as seed, usually if no other opportunity exists. The formal institutional system typically reflects a one-directional flow of material: the material is originally derived from the local farmers' systems. It is stored in genebanks which principally serve breeders. Breeders recombine genetic diversity and select a portion of this, which thereafter is multiplied and diffused by seed-producing organizations or programmes. [Adapted from de Boef *et al.* (1997) and Almekinders and Louwaars (1999).]

seed banks requires substantial coordination and training. A similar reasoning will ask breeders to involve grassroots organizations or extensionists to facilitate the participatory breeding activities when these increase to a scale beyond the breeders' capacities to deal with all involved farmers and experiments. Collaboration with organizations that can cover seed production and diffusion in an up-scaled situation is essential as well. The collaboration in up-scaled, mainstreamed situations will be more than that of participating farmers. Integrated approaches will involve other organizations which play roles in the PGR system as well. Thereby not only will the linkages with the farmers' systems be developed, but also inter-institutional linkages, including those with NGOs.

When these approaches do indeed become mainstream, the picture of the PGR system will not look like the one in Figure 2.1 with formal and informal systems. These two systems will then be integrated, with multiple linkages between them. In such an integrated PGR system, farmers will play a role in conservation, breeding and seed supply which is recognized and supported by the genebanks, plant breeders, seed producers and NGOs or other extension agencies.

Conclusion

When participatory approaches – of which the cases in this book are examples – are successfully integrated into PGR institutions, farmers will have better access to suitable diversity for further development. Having more sustainable systems, farmers may have better opportunities to continue using genetic diversity as a tool in facing the environmental variation and socioeconomic uncertainties. This also provides the rationale for the conservationists to support the breeders, seed producers and other development-oriented organizations. Improved access to genetic diversity will have a beneficial effect on both development and on-farm employment of diversity. Considering on-farm employment of diversity as *in situ* conservation of genetic recourses may seem too easy and not deserving the label 'conservation' at all. It may, however, represent a dynamic view of *in situ* conservation which contributes to more sustainable agriculture in marginal areas and thereby to farming systems which will continue to have the capacity to maintain and develop genetic resources for present and future use.

References

Almekinders, C.J.M. and N.P. Louwaars. 1999. Farmers' seed production. New approaches and practices. Intermediate Technology Publications, London.

Byerlee, D. 1994. Modern varieties, productivity, and sustainability: recent experiences and emerging challenges. CIMMYT, Mexico.

Clawson, D.L. 1985. Harvest security and intraspecific diversity in traditional tropical agriculture. Econ. Bot. 39:56-67.

Boef, W. de, J. Hardon and N.P. Louwaars. 1997. Integrated organisation of institutional crop development as a system to maintain and stimulate the utilisation of agro-biodiversity at the farm level. Paper presented at the International meeting Managing Plant Genetic Resources in the African Savannah, Bamako Mali, 24-28 February 1997.

Hardon, J.J. and W.S. de Boef. 1993. Linking farmers and breeders in local crop development. Pp. 64-71 *in* Cultivating Knowledge. Genetic diversity, farmer experimentation and crop research (W. de Boef, K. Amanor, K. Wellard and A. Bebbington, eds.). Intermediate Technology Publications, London.

Prain, G. and J. Hagmann. 2000. Farmers' management of diversity in local systems. Pp. 94-100 *In* Encouraging Diversity. The synthesis between crop conservation and development (C.J.M. Almekinders and W.S. de Boef, eds.). Intermediate Technology Publications, London.

3. A brief review of participatory tools and techniques for the conservation and use of plant genetic resources

Amanda King

One of the defining characteristics of participatory methodologies is the use of flexible research tools, and ways of eliciting information that are accessible and agreeable to those participating. Outlined below is a basic list of tools that may be used for participatory research on plant genetic resources. They are described in a sequence of how they may be used in the field, beginning with the collection of baseline data to guide sampling, to the carrying out of more extensive and in-depth studies. This should not be interpreted as a rigid sequence; many tools are useful at several stages of research. Some tools may be used concurrently. None of the following tools are necessarily participatory; the ways in which the tools are adapted and carried out with the inclusion of and facilitation by community members are what define them as participatory tools.

Developing a sampling procedure

Units of analysis

An important initial consideration for collecting of sociocultural data and farmer knowledge is the level of aggregation and the units of analysis to be used. While the 'household' is often used as a key point of reference, the definition of household varies according to cultural context. In order to use the household as a basic unit of research, it is necessary to clearly define what is meant by a 'household' in a particular community, and to analyze it as both a productive and social unit. Using the 'household' as a unit of analysis may hide disparities of knowledge, experience and power among individuals within it. At the same time, focusing on individuals alone diminishes the important social and cultural dynamics that take place between household members.

In order that all types of community knowledge are represented in participatory research, it is necessary to look at both inter-household and intra-household variables (Hardon-Baars 1997). Household crop production and farmer decision-making may be influenced by inter-household factors such as the land tenure system or the size of land holdings. In addition, crop management may be shaped by factors within the household such as differential access to inputs, responsibility and control over products. In order to capture information about responsibility and ownership, as well as differences in use patterns and value systems, it is necessary to collect data at both the individual as well as the household level. In addition, the concurrent use of gender analysis will help to reveal differences in management decisions, responsibilities and values that may otherwise remain hidden.

Sampling issues

Because properties, management and uses of the crop material are of concern in research on genetic diversity, research sampling strategies should also take into account ecological, land management and seed/plant management patterns. Accordingly, plant/seed sampling and social sampling must be linked in such a way as to establish the connections between social and biological factors, and patterns in genetic diversity. A key source on the various aspects of sampling plant populations and collecting materials is Guarino *et al.* (1995), while information on technical aspects of conducting social science research can be found in Frankfort-Nachmias and Nachmias (1996). The combination of scientific and social data and the integration of different types of sampling strategies also has been discussed (Jarvis and Hodgkin 1998). However, there is a need for continued experimentation and adaptation of methods.

The development of a sampling strategy may start with guided samples based on a small number of households, which can be used to establish the key variables for further study. Methods used in guided sampling include structured surveys, key person interviews, group interviews, focus groups, and more creative and participatory ways of eliciting information. Once this initial information is collected, it is possible to conduct more extensive research. Selection of households for more detailed data collecting may take place on either a random or a directed basis.

Various types of sampling strategies include:
- random sampling – the selection of households or individuals on a random basis
- stratified random sampling – groups or strata of the population are separated for certain features (for instance people with land and landless people), each group/strata is treated as a separate case, and a sample established for each
- cluster sampling – individuals or households are chosen in groups or clusters and not on an individual basis, and within each cluster a random sampling method is used (for example, one cluster may be those individuals who plant in a dry area with poor growing conditions)
- multistage sampling – samples are selected using simple random sampling, and from these samples, a new set of samples is drawn (Davis-Case 1990).

Collecting of available information

Collecting all relevant information should be the first step in any research process because it saves time and duplication of effort, and because it prepares the researcher for interactions with the community. Previously published information which may be useful for researching crop diversity relates to crop ecosystems, the communities that manage them, or to the crops themselves. Existing environmental data, ecological and geographical maps, as well as social or anthropological studies are particularly helpful (Guarino *et al.* 1995).

Participant observation

Definition: Participant observation is a classical anthropological tool which has been used predominantly to study community and individual behavioural patterns. Participant observation is the process of documenting observations in a systematic and continuous way, without disrupting the processes, people or locations being observed.

Uses: Participant observation can be used to gain baseline information about human communities, behavioural or management patterns, as well as social structures and human interactions. In addition, basic observation can be used to assess crop populations, diversity among crops, phenotypic variation and crop ecosystems. This information can be used to develop a sampling strategy for more in-depth research, or it can be used to support other types of data. In order to be genuinely participatory, participant observation must be adapted so that the process of observation and the documentation of information are not solely the responsibility of the researcher. Ways in which to share these tasks might be that participants monitor their own activities, or the activities of those close to them, with the corroboration of other community members. This may also take the form of daily diaries, discussions of daily activities, or the observation of activities or sites by teams made up of paired researchers and community members.

Advantages: Basic observation is the simplest way of obtaining a general understanding of the variables that are going to be researched. When carried to more extensive levels, observation can provide the researcher with a great deal of easily accessible and highly useful information.

Disadvantages: Participant observation is frequently not the neutral tool that it is intended to be; the mere process of observing often influences the subject that is being observed, and the data that result may be biased by the individual interpretations of the researcher. In addition, participant observation is not an interactive form of research, unless members of the community act as both the observers and the observed. As discussed previously, the format of participant observation may be altered with the framework of participatory research so that the activity may be more inclusive to the perceptions of community members.

Participatory Rural Appraisal (PRA)

Definition: This is an intensive, iterative and expeditious form of research, which relies on small multidisciplinary teams that employ a range of methods, tools and techniques specifically selected to enhance understanding of rural conditions by tapping the knowledge of local inhabitants. Its outstanding characteristics are flexibility, minimal resource requirements, and the central role given to intensive dialogue, varied types of communication, and researcher-community cooperation in order to access community knowledge. Triangulation is a common technique employed in choosing methods, sites and participants in research, so that a minimum of three perspectives provides a range of variables to be studied.

PRA places a strong emphasis on sharing ownership with participating communities, through the incorporation of community goals and needs into the design, objectives and uses of the research. With the new questions and insights generated by conducting basic exercises with communities, researchers can move more directly toward understanding problems and facilitating the development of appropriate solutions. Modifications of previous methods, as well as new tools for this type of research, are constantly being generated as researchers develop their own means of working interactively with communities. A partial list of commonly used tools is given below. Most of these tools are effective in eliciting the specialist knowledge related to gender or other factors when conducted with separate focus groups.

Uses: PRA in particular can provide useful tools for conducting various types of participatory research. PRA techniques can be used to gain both a general and a more in-depth understanding of community knowledge. A general understanding of community characteristics can help to guide the development of a sampling strategy for further research, while more extensive community knowledge can be used to supplement other types of qualitative and quantitative data.

Advantages: Most of the PRA techniques are designed to be inexpensive and easy for anyone to participate in. They generate a great deal of information in a short time and provide insight into social behaviours and management practices.

Disadvantages: PRA techniques require a capable and experienced facilitator. Interpreted out of context and taken on their own, the data produced from these techniques can be superficial. They should be used in conjunction with other tools as a means to generate new perspectives and research orientation. Special care must be taken to ensure that these exercises are carried out in a participatory manner, and that community members are involved in adapting tools to suit their own contexts.

One difficulty in using these tools is that they may have the effect of altering the cultural mode of expression. According to Cornwall and Jewkes (1995: 1673), the very act of the 'community' engaging with outsiders necessitates a simplification of their shared experiences into a form and generality which is intelligible to an outsider. Even though they are designed to highlight areas of diversity among participants, these tools may continue to mask differences between individuals within a group, and between the knowledge systems of researcher and participant.

Partial list of PRA tools
- Social resource/ Historical mapping (Fig. 3.1)
- Transect maps (Fig. 3.2)
- Preference matrix ranking
- Rating, sorting exercises
- Semi-structured interviews
- Local knowledge forms – folk taxonomies
- Seasonal calendars (Fig. 3.3)
- Labour calendars (Fig. 3.4)
- Logic/Decision trees (Fig. 3.5)
- Drawing – bar, venn, flow diagrams (Fig. 3.6, Fig. 3.7)
- Leadership Communication Network/ Farmer Network Analysis (Fig. 3.8).

Brief tool descriptions

Community mapping/Historical mapping

Study of resource management requires knowledge of both the spatial distribution of resources and of how these resources are utilized. These exercises involve the community in mapping their resource base in order to generate information about the local environment and social systems, gauging community perceptions of ownership, responsibility, physical or social boundaries, and clarifying relationships between environmental factors and agricultural activities.

Using previously drawn maps, participants can identify the exact location of resources and patterns of resource usage. Participants can themselves map local infrastructure, land tenure systems, spatial distribution of crops and their relationship to natural resources (Fig. 3.1). It is useful to do mapping in the field so that it can be supported by direct observation. The information generated from mapping can be used to develop a sampling strategy or to collect detailed data. Mapping also helps to build rapport with the farming community.

Historical mapping can be used to document the history of the community or a certain groups within the community, and can be done in pictures, writing or symbols. The timetable may be focused on a specific subject such as natural or communal resource management, or the impact of village growth or economic change on the surrounding environment. This tool can give a temporal dimension to studies of diversity.

Transect walks

The purpose of transect walks is to provide a good representation of the social or biological variation within an area being studied, as well as to document as much information as possible from direct observation of the community and the local environment. One use of a transect walk is to delimit the main agro-ecological zones within a community, chosen subjectively as being distinct in terms of one or more ecological, agricultural, social or economic feature. Another use may be to illustrate the variation and spatial location of social units found within a community, in order to develop an appropriate sampling strategy (Fig. 3.2). Transect walks are particularly useful for identifying the small niches within the agricultural landscape that may contain pockets of diversity.

Ranking, rating and sorting exercises

These tools are simple and inexpensive ways to provide insight into individual or group decision-making and to identify the criteria that people use to select certain items or activities. When used with different groups and compared, they can pinpoint differences in perception, identify priorities and monitor changes in preference. In addition, they can translate qualitative information into a quantitative form. This type of information is valuable for understanding the ways in which communities value and manage crop species.

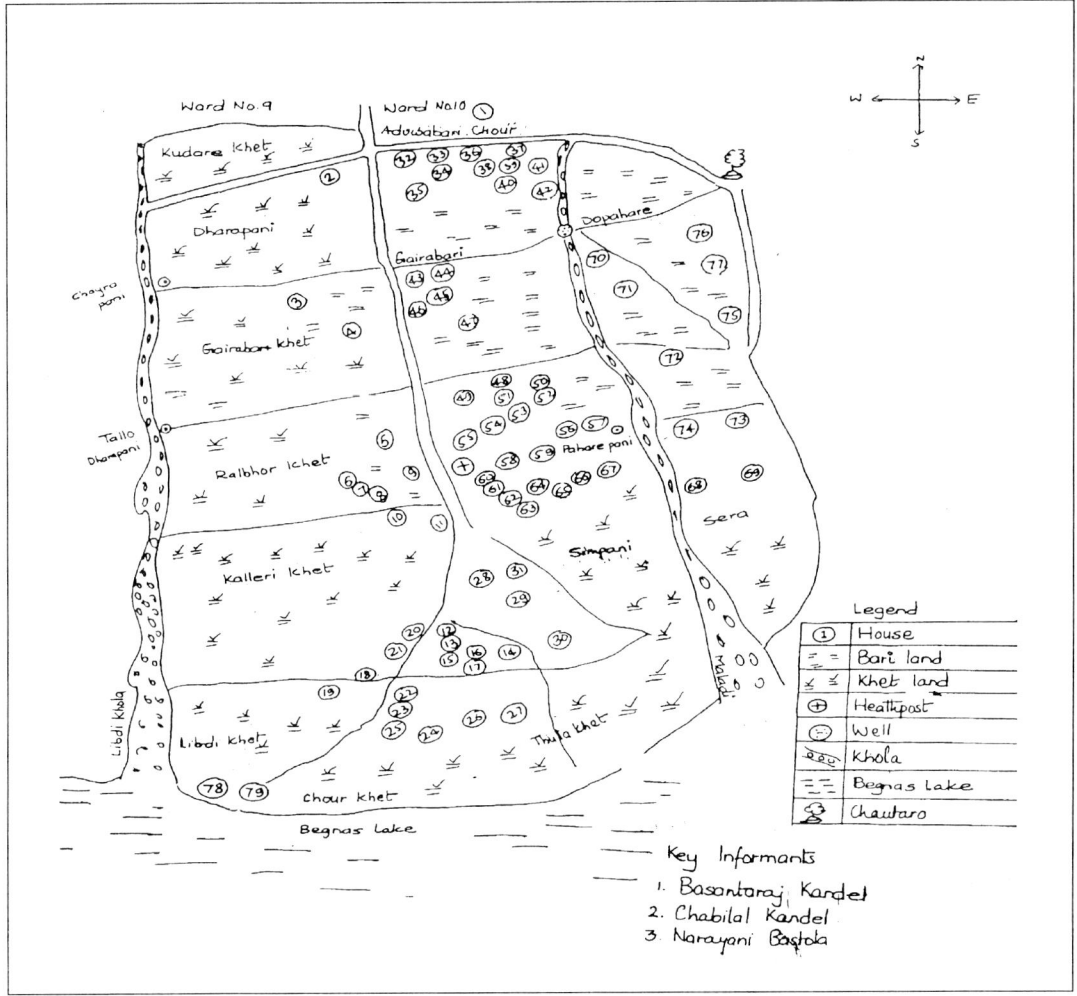

Fig. 3.1. Social/resource map of Lekhnath Ward Nos. and 10 (Source: Rijal, unpublished)

Ranking – The process of ranking a certain number of items on the basis of a certain criteria. For instance, participants might rank tree species on the basis of their general usefulness, where usefulness is defined by group criteria. Preference ranking or preference matrix ranking tools are a useful way of understanding preferred traits in a variety and used to set breeding goals in participatory breeding. A detailed instructional unit on reference ranking was published by CIAT (Guerrero et al. 1993).

Rating – This process, which works best with literate people, involves rating certain statements or ideas on a scale which runs from complete agreement to total disagreement. For example participants may be given a statement about a method of crop management and asked to rate how strongly they agree with the statement.

Sorting – The process of sorting a unit according to its characteristics into clearly defined categories. For instance, participants could sort households between three categories of household economics. The defining characteristics of each category can be decided by the group.

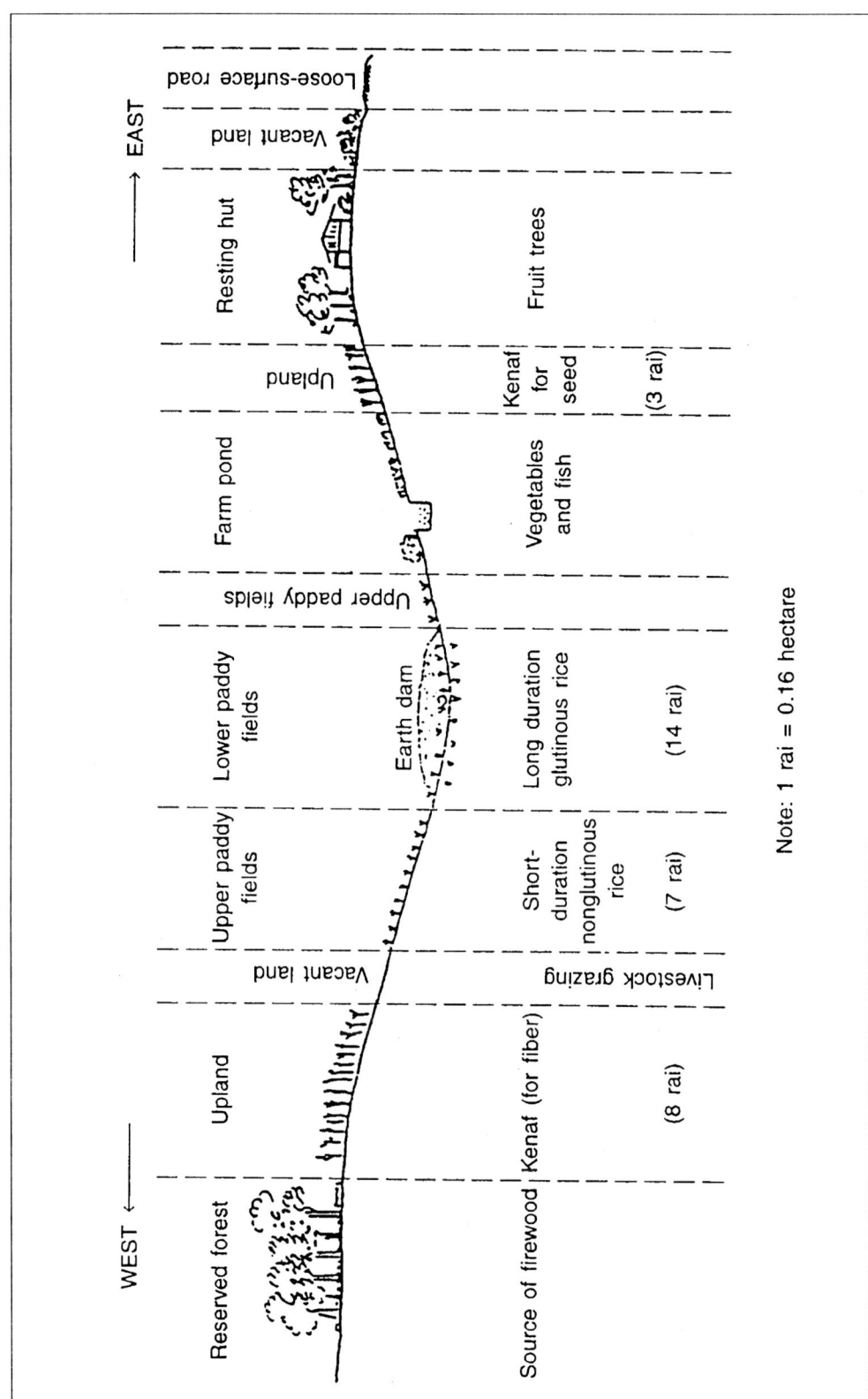

Fig. 3.2. Transect of the undulating farm land of a Khon Kaen Farm household (Source: KKU-USAID Farming Systems Research project, unpublished)

Semi-structured interview

Semi-structured interviews are interviews conducted with individuals or groups, focused on a particular issue. While an interviewer may have a checklist of information to cover, interview questions are not rigidly structured and may be adapted according to the directions that responses take. In other words, the interaction is based upon an open framework which allows for focused, conversational, two-way communication. This type of interview is useful because it allows researchers to obtain specific quantitative and qualitative information from a sample of the population, to probe for unknown information, and to get a broad range of insights. For more information on interviewing, see below.

Local knowledge – terminology

One method of assessing the diversity preserved in agro-ecological systems is by determining the specific values that individuals assign to crops, and the reasons these crops continue to be grown and used within the community. Determining the ways in which farmers perceive certain varieties to be distinct can be accomplished by questioning farmers about the distinctive uses of the crops, the variation they perceive in crop properties, and about the names they give to different varieties. In linguistically complex regions where different languages and dialects are found, recognizing and understanding local terminology is important. Using the local terms when asking questions helps interviewers to gather accurate information. In addition, folk taxonomy can be used as a tool to understand how people classify and value resources and environments, which in turn is reflected in their different strategies of management.

Seasonal calendars/Labour calendars – activity sequences

Preparing seasonal calendars with communities which outline an entire agricultural season, the crop sequences grown and associated tasks can supply information on environmental factors, as well as management decisions, value systems and labour responsibilities (Fig. 3.3).

Labour calendars are a similar tool which focus on the labour tasks performed throughout the agricultural season. This tool is an especially useful tool for illustrating gender-differentiated responsibilities and management of crops (Fig. 3.4). If particular farm work can be broken down into an activity sequence, it may be informative to ask questions about the individual activities. Separate activity sequences can be determined for years where environmental or community conditions may have altered, in order to assess the impact of such changes.

Logic/Decision trees

Decision trees can be used to identify distinctive livelihood systems, farmer strategies and decision-making which shape the management of crop diversity. Decision trees may be constructed from the information gathered in transect walks, through direct observation and through interviews. The logic tree can be drawn to classify farmers by types of operation or pattern of resource usage as observed by the researcher. The diagram of the logic tree should include key determinants placed at strategic branching points. The decision tree can be drawn to illustrate the key factors or important conditions that make the farmer decide to adopt one type of cropping pattern or management in preference to others (Fig. 3.5).

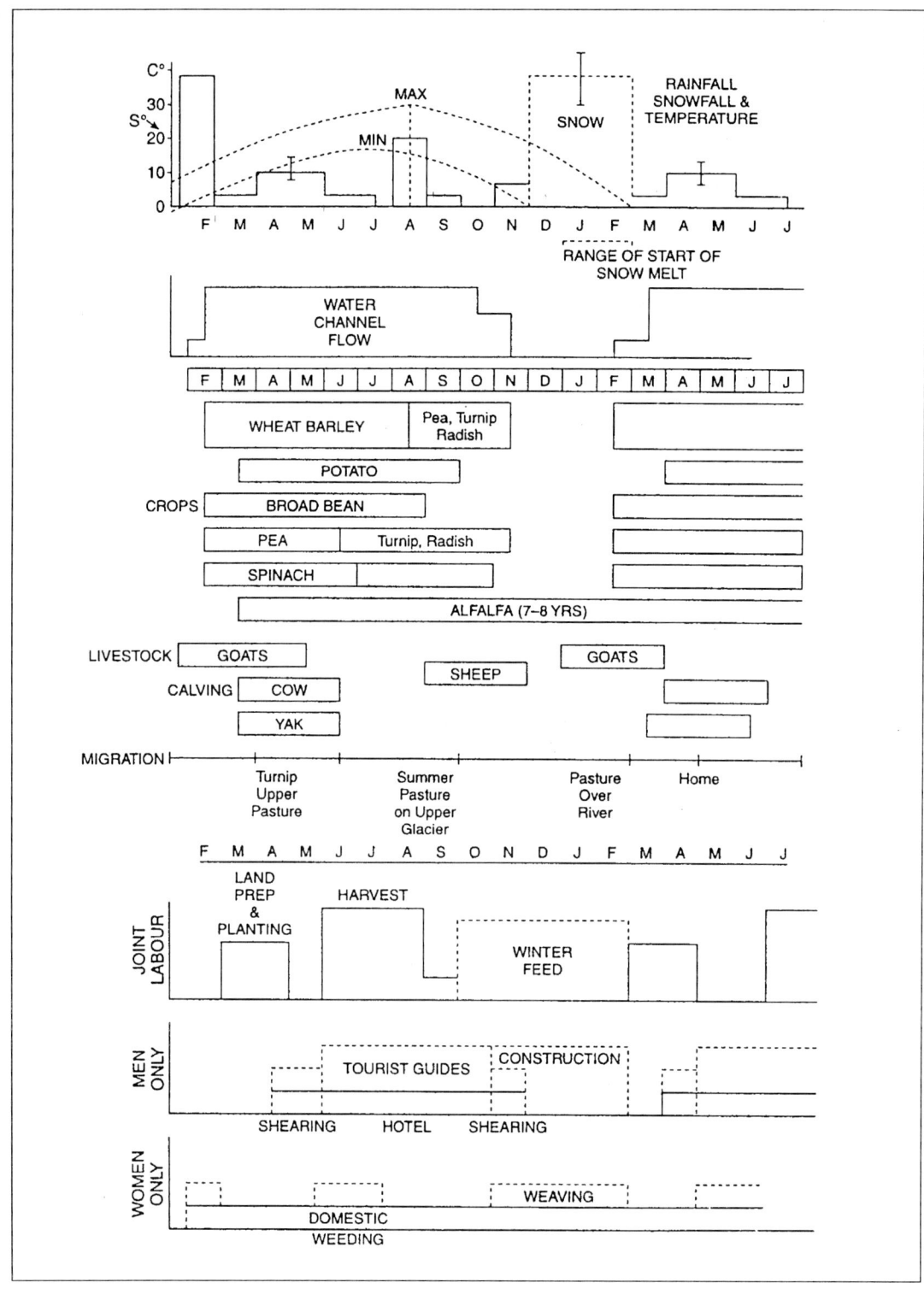

Fig. 3.3. Seasonal calendar for Passu (Source: Conway *et al.* 1987).

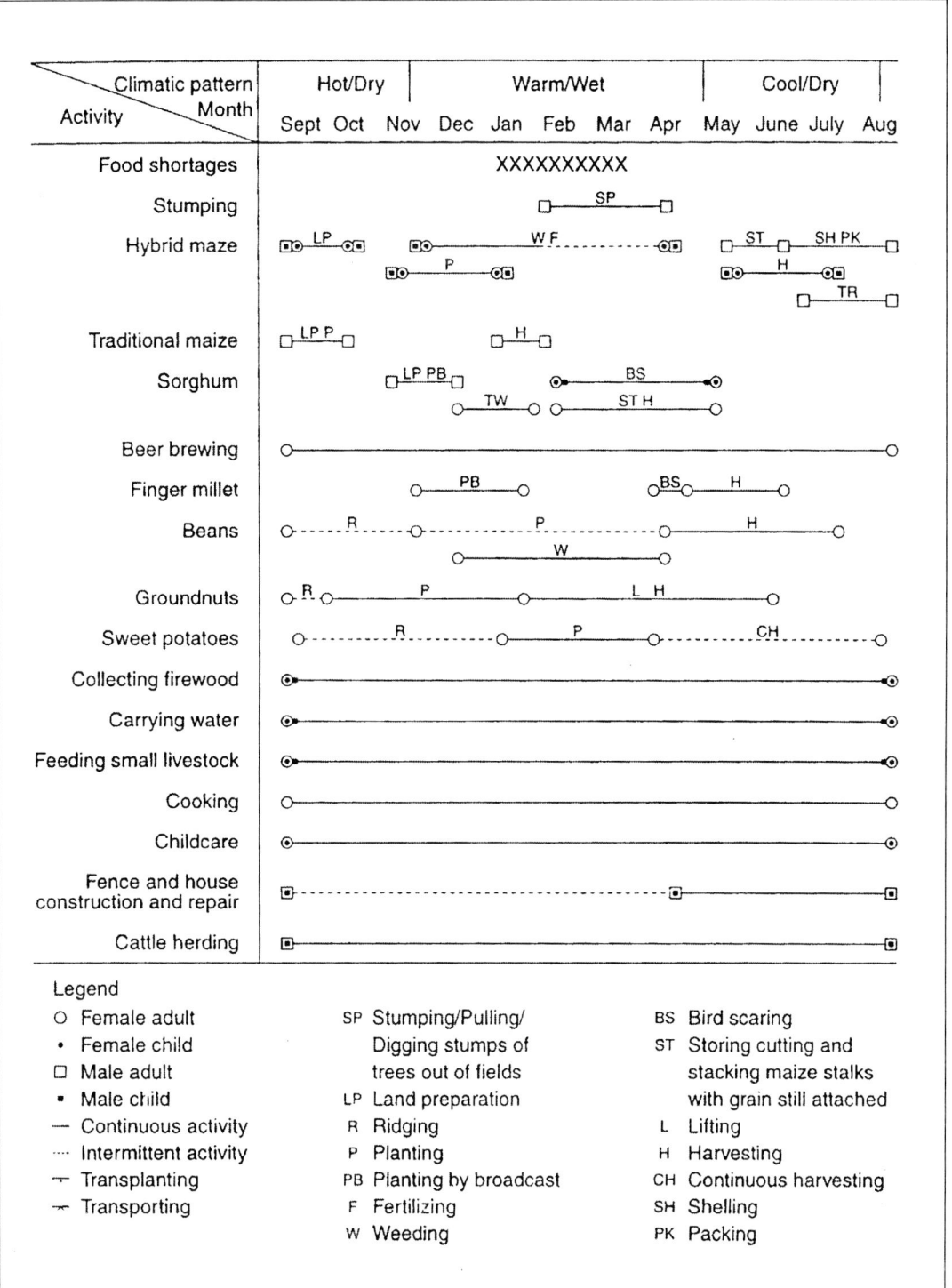

Fig. 3.4. Gender-disaggregated activity calendar for Mkushi District (Source: Feldstein and Poats 1989).

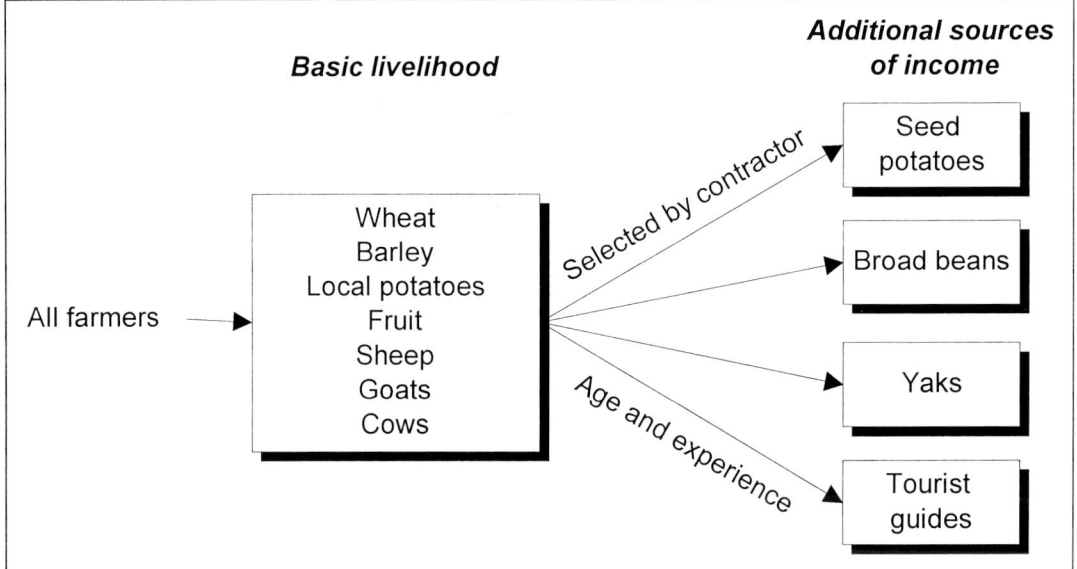

Fig. 3.5. Decision Tree for livelihood systems in Passu (Source: Conway *et al.* 1987).

Diagrams

Structured diagrams are a tool for illustrating farmers' knowledge in a quantified or conceptual way. Bar diagrams, flow diagrams and venn diagrams can all be used to illustrate different conceptual properties. Bar diagrams illustrate proportional relationships, such as the proportion of different resources held by different types of farmers. Flow diagrams are designed to show the interrelationships between different variables, such as the interrelationship between production and marketing and the costs and returns at different stages (Fig. 3.6). Venn diagrams are also a way of showing interrelationships, often between institutions or groups of decision-makers (Fig. 3.7).

Drawings and diagrams are also a useful tool for eliciting farmer knowledge, especially in the case in which community members are oriented toward visual forms of expression, or do not share similar languages. Furthermore, while each gender knows its role within the production system, this is often in an implicit rather than an explicit manner. A diagram can help record and reflect this knowledge and provide a course for further reflection.

Interviews – Group/Focus Group/Individual

Definition: A group interview is a gathering of people with a facilitator for discussion of an issue. The meeting can involve a large number of people or a smaller number who focus on a specific problem or purpose. Semi-structured interviews are those in which there is a specific agenda to be discussed, but there remains a degree of flexibility. This ensures that the individuals discussing the issue are able to modify the direction of the interview according to the information that is revealed.

Uses: Meetings can be used to gather general and commonly shared information. For example, questions may be asked about the community structure and function, the characteristics of local ecosystems, commonly held natural and agricultural resources, the predominant crop varieties raised by the community, the types of commonly encountered pests or difficulties with certain varieties, commonly held perceptions of the uses and values of particular crops, etc.

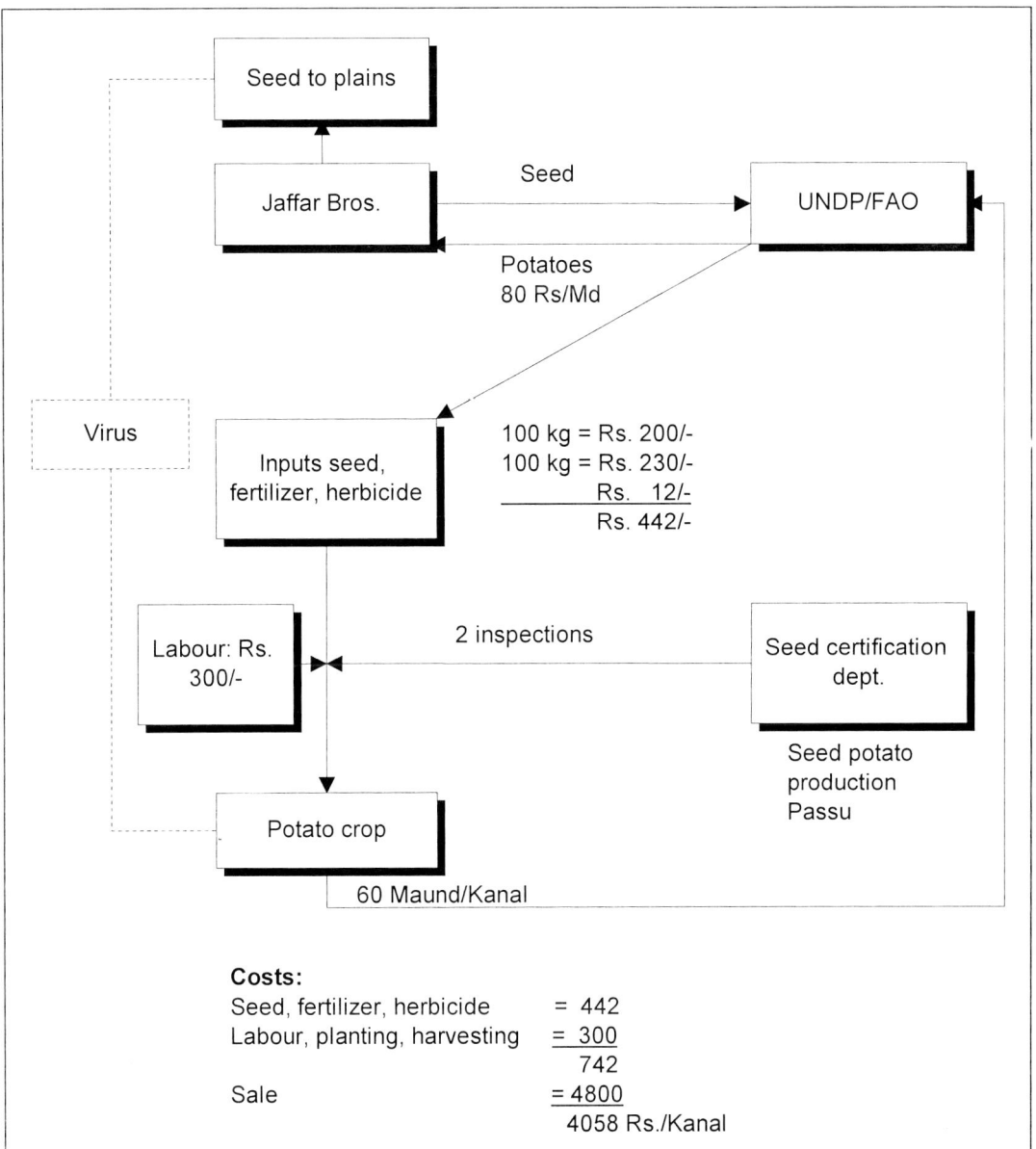

Fig. 3.6. Flow diagram of seed potato production and marketing in Passu (Source: Conway *et al.* 1987).

Advantages: Meetings are useful in that they reach many people in a short time, they elicit commonly shared information and encourage a flow of ideas between group members, and they help to establish a rapport between researchers and community members.

Disadvantages: One disadvantage of meetings is that while they allow the sharing of common information, the views and specialized knowledge of certain individuals or marginalized groups are commonly not heard.

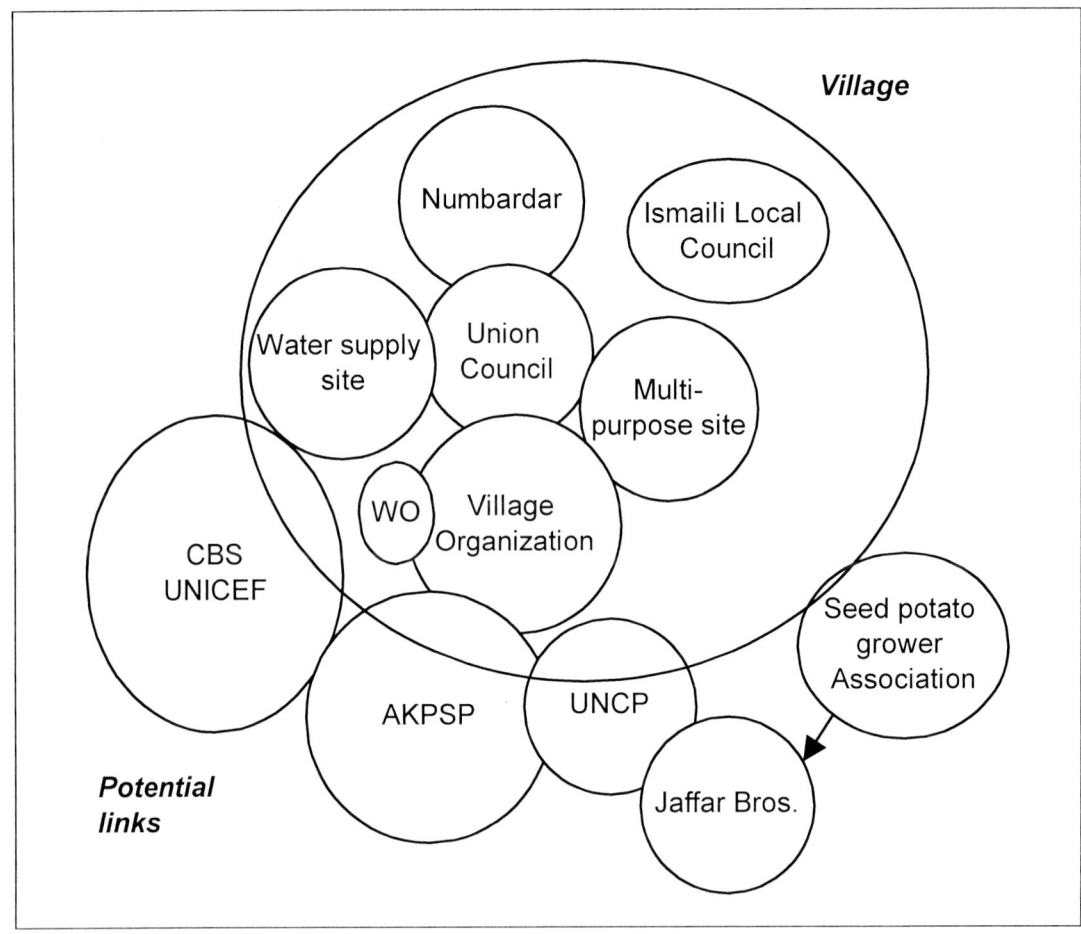

Fig. 3.7. Venn diagram of institutional overlap in Passu (WO = women's organization) (Source: Conway et al. 1987).

Focus group interviews

Definition: Focus group meetings are made up of people with similar concerns, who can speak comfortably together, and who share a common problem and purpose. Focus group meetings can be used as a tool to elicit knowledge shared by a certain group which is not expressed in the context of a larger gathering. This information can be compared with that generated by the larger group.

Uses: Focus groups can be used to generate information that is shared among a smaller group of people. Usually all the questions that are raised with larger groups can be raised in the context of focus group interviews in order to obtain the specific views and specialized knowledge of the group being addressed. For example focus groups can be asked questions about what are the major uses for a particular species *by a particular group*, how is the species managed, who is it managed by, and why the species is perceived as useful. Comparing responses made by smaller and larger groups may generate insights about the individual groups represented, as well as the interaction between different factions of the community.

Advantages: Views and specialized knowledge which are not expressed within larger groups can be elicited.

Disadvantages: No matter how small the group, there is still a tendency for some individuals to dominate the discussion. To obtain the knowledge of all group members, it may be necessary to conduct personal interviews, or to use questionnaires.

Individual interviews

Definition: These are interviews conducted with one informant in order to elicit the specific knowledge of the individual. Key informant interviews are interviews with individuals who are particularly knowledgeable about a particular issue, who are accessible and who are willing to talk.

Uses: Individual interviews may generate any of the types of information described above, and are particularly useful for eliciting quite specific, individually held information. They also can be used to ascertain unique views not presented elsewhere.

Advantages: These interviews are least influenced by the physical presence of other members of the community. They are the most direct way of understanding individual knowledge and management patterns.

Disadvantages: To be of use when studying a community or a crop population, many individual interviews must be conducted in order to elicit the knowledge and information of a group. Individual interviews are costly in terms of time and other resources.

- Have a clear purpose for the meeting and develop an agenda which includes researcher and community goals.
- Obtain the approval and involvement of local leaders. Be aware of local customs and protocol.
- Arrange a convenient time and place for the meeting, considering both the size and composition of the group
- Select a practised facilitator, and plan a strategy to encourage discussions and two-way communication.
- Hold separate focus group meetings for factions of the community who are unable or unwilling to speak up in larger gatherings.
  (Adapted from FAO 1998)

Questionnaires

Definition: Questionnaires are lists of questions designed to elicit specific information from individuals or from the primary research samples being studied within a community, e.g. households, groups working on the same agricultural plot, etc. They are usually used with selected samples that have been chosen out of the entire population by means of a rough characterization tool, such as focus groups. Questionnaires gather both quantitative and/or qualitative information. While they may be in the form of a survey which the participant fills out, they are usually a series of questions delivered orally by a researcher who then records the individual responses. Data from questionnaires are pooled and may be analyzed in order to obtain information and statistics related to specific issues.

Uses: Questionnaires may be used to gather specific quantitative and qualitative data from a research sample which can be used to support hypotheses, or to explore relationships between variables. These data can be about individuals, households, parcel/plots, communities or ecosystems. Quantitative questions directed to households include questions such as what is the household composition, gender composition, ethnicity, living standard, tenure, educational status, etc. Questions about parcel/plots may include the land quality, purchased inputs, labour responsibilities, use of crops, seed source, perceived genetic diversity, etc. Qualitative questions on questionnaires may take the form of a ranking exercise such as the ranking of the values perceived in a crop. This type of information *should not replace* other types of qualitative information, but should be used in conjunction with other tools, in order to obtain a more holistic picture of the issue being researched.

Advantages: Questionnaires allow the translation of individual knowledge into a quantitative format. This quantification is valuable because it can be used to measure certain characteristics, to explore the relationship between variables, to gain a statistical understanding of a community or crop population, and to argue for or against hypotheses about a communities' maintenance and use of diversity.

Disadvantages: The most useful questionnaires are precise and well-honed tools. To work efficiently they must be used with a well-defined sample to explore a specific issue. They are often quite long and complex, because of the amount and detail of the information being sought. It is essential to keep questionnaires both relevant and concise so that they do not become a burden to the participant or to the researcher. In addition, questionnaires are the least interactive form of information retrieval; they do not typically allow for any reciprocal exchange of knowledge or input from participants in the research process. One solution to this problem is to solicit the help of the community in designing and administering the questionnaire. However, questionnaires should not play too great a role in a participatory research methodology, because of the existence of other, more interactive and empowering ways of obtaining information.

Farmer network analysis

A farmer network in a social system refers to the inter-relationship among a set of individuals for seeking information and genetic resources and exchange. Understanding and examination of such network relationships provide significant insights about social and individual behaviour, and the process by which innovations are developed and disseminated among people in a social system (Fig. 3.8).

Further information on tools for the elicitation of farmer knowledge

The final part of this paper consists of lists of sources where information can be obtained on the use of specific tools for participatory methodologies. This list is adapted from Harrington (1997).

Assessment of Local Knowledge Systems: Folk taxonomies, farmer classification of land types, traditional systems of organization, oral histories, status distinctions, decision point analysis (e.g. Warren and Cashman 1988; Sandoval-Nazarea 1994).

Interview Techniques: semi-structured surveys, key informant interviews, the use of focus groups, individual interviews (e.g. Byerlee and Collinson 1980; Beebe 1985; Slim and Thompson 1993)

Community Exploration Techniques: Community appraisals, group treks, participatory workshops, rapid site description, transects, biophysical assessments, indigenous indicators (e.g. Chambers and Ghildyal 1985; Conway *et al.* 1987)

Rapid Rural Appraisal: (e.g. FAO 1998; Khon Kaen University 1987; McCracken *et al.* 1988)

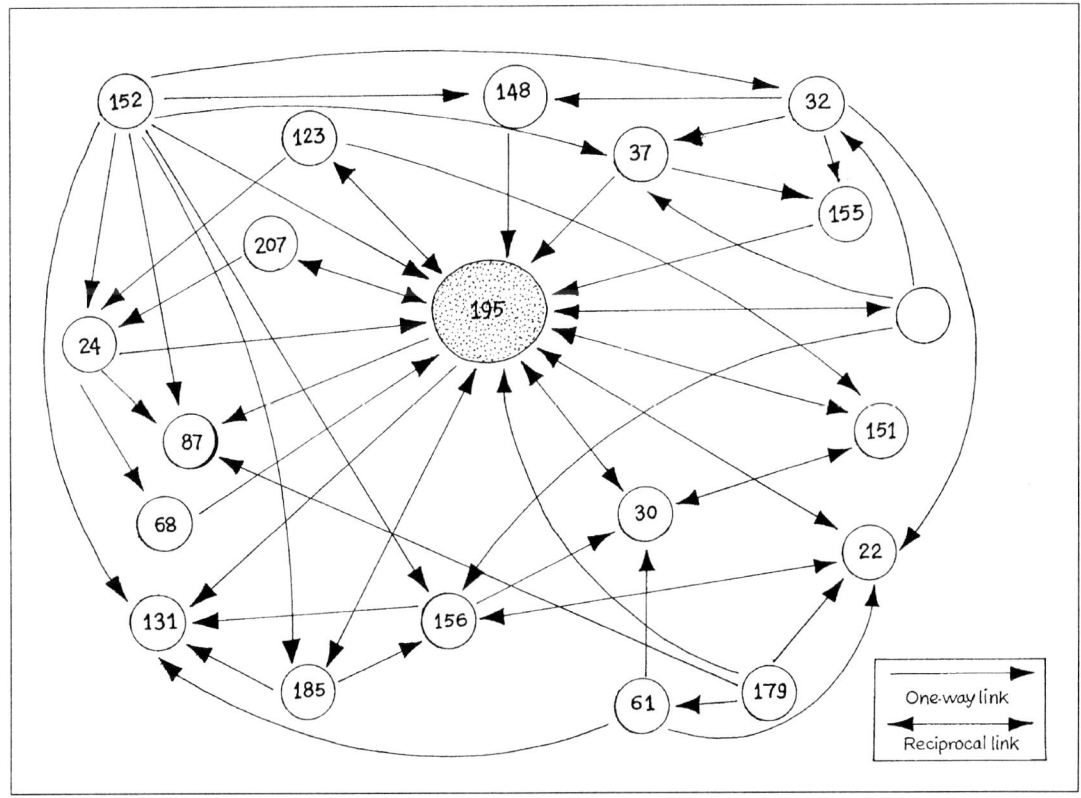

Fig. 3.8. Leadership Communication Network (Subedi and Garforth 1996).

Participatory Rural Appraisal: (e.g. Schönhuth and Kievelitz 1994)

Diagramming Techniques: resource flow diagrams, seasonal diagrams, decision trees, problem-cause diagrams (e.g. Lightfoot *et al.* 1989)

Mapping Techniques: sketches, historical patterns, agro-ecosystem zoning
(e.g. Chambers 1990)

Time Flow Analysis: seasonal calendars, time lines, time allocation studies
(e.g. Maxwell 1984; Triomphe 1995)

Farmer Experimentation: farmer's adaptations, farmer-managed experiments, farmer selection from among multiple alternatives
(e.g. Ashby *et al.* 1987; Quiros *et al.* 1991; Prain *et al.* 1994)

Preference Ranking: (e.g. Guerrero *et al.* 1993).

Conclusion

Multiple benefits are generated by the participation of farmers in research on crop genetic diversity. Designing research in order to meet local needs and values, and to build the capacity of communities, increases the utility of both the process of researching and the products of the research. While farmer participation is desirable because of the value it imputes to research, there are many ethical issues that arise from participation in research, including questions of ethical conduct in work with local communities, issues of ownership over genetic resources and intellectual property, and finally the importance of benefit-sharing and the development of equitable systems of resource exchange. The ethical implications of participatory research will necessitate increased care over the ways in which community and individual knowledge are handled in a public arena, in addition to the development of protocols for participatory research, and systems of resource protection and exchange that are acceptable at global, national and local levels. Finally, effective and truly participatory research tools also empower communities in several ways. First, the use of participatory research tools helps communities to better understand their situation and identify their own solutions. Second, use of the tools can help communities to communicate more effectively with outside agencies and influence decisions affecting their livelihoods and environment. Third, participatory tools affirm the validity of local knowledge and place it on a more equal footing with other knowledge systems.

References

Ashby, J., C.A. Quiros and Y.M. Rivera. 1987. Farmer Participation in On-Farm Varietal Trials. Network Discussion Paper 22. Overseas Development Institute Agricultural Administration (Research and Extension), London.

Beebe, J. 1985. Rapid Rural Appraisal: The Critical First Step in a Farming Systems Approach to Research. Networking Paper No. 5. Farming Systems Support Project, Washington, D.C.

Byerlee, D. and M. Collinson. 1980. Planning Technologies Appropriate to Farmers; Concepts and Procedures. CIMMYT Information Bulletin. CIMMYT, Mexico, D.F.

Chambers, R. 1990. Microenvironments Unobserved. Environment and Development Gatekeeper Series No. 22. IIED, London.

Chambers, R. and B. Ghildyal. 1985. Agricultural Research for Resource-Poor Farmers: the Farmer-First-and-Last Model. Discussion Paper No. 203. Institute of Development Studies, Brighton.

Conway, G.R., J.A. McCracken and J.N. Pretty. 1987. Training Notes for Agro-ecosystem Analysis and Rapid Rural Appraisal. International Institute for Environment and Development, London.

Cornwall, A. and R. Jewkes. 1995. What is Participatory Research? Soc. Sci. Med. 41(12):1667-76.

Davis-Case, D. 1990. The Community's Toolbox: The Idea, Methods and Tools for Participatory Assessment, Monitoring and Evaluation in Community Forestry. Community Forestry Field Manual 2. Food and Agriculture Organisation, Rome.

FAO. 1998. Socioeconomic and Gender Analysis (SEAGA): Field Handbook. Food and Agriculture Organisation, Rome.

Feldstein, H.S. and S.V. Poats, editors. 1989. Working Together: Gender Analysis in Agriculture. Vol. 2. Kumarian Press, Inc., West Hartford, Conn., USA.

Guarino, L., R. Rao and P. Reid. 1995. Collecting Plant Genetic Diversity: Technical Guidelines. CAB International, Wallingford.

Guerrero, M. del P., J.A. Ashby and T. Garcia. 1993. Farmer evaluations of Technology: Preference ranking. Instructional Unit No. 2. CIAT, Cali.

Harrington, L. 1997. Doctors, lawyers and citizens: Farmer participation and research on natural resources management. Pp. 53-63 *in* New Frontiers in Participatory Research and Gender Analysis. CIAT, Cali, Colombia.

Khon Kaen University. 1987. Proceedings of the 1985 International Conference on Rapid Rural Appraisal. Khon Kaen University, Thailand.

Lightfoot, C., C. Axinn, P. Singh, A. Bottrall and G. Conway. 1989. Training Resources Book for Agro-Ecosystem Mapping. IRRI, Phillipines.

Maxwell, S. 1987. Hitting a Moving Target; II. The Social Scientist in Farming Systems Research. Agriculture and Rural Problems Discussion Paper No. 199. Institute of Development Studies,

Brighton.

McCracken, J., J.N. Pretty and G.R. Conway. 1988. An Introduction to Rapid Rural Appraisal for Agricultural Development. International Institute for Environment and Development, London.

Prain, G., H. Fano and C. Fonseca. 1994. Involving Farmers in Crop Variety Evaluation and Selection. UPWARD Training Document Series 1994-2. UPWARD, Los Banos, Laguna.

Quiros, C.A., T. Garcia and J. Ashby. 1991. Farmer Evaluations of Technology:Methodology for Open-Ended Evaluation. Instructional Unit No. 1. IPRA Project/CIAT, Cali.

Sandoval-Nazarea, V. 1994. Memory Banking Protocol: A Guide for Documenting Indigenous Knowledge Associated with Traditional Crop Varieties. Training Document Series 1994-2. UPWARD, Los Banos, Laguna.

Schönhuth, M. and Z. Kievelitz. 1994. Participatory Learning Approaches: Rapid Rural Appraisal, Participatory Appraisal, An Introductory Guide. TZ-Verlagsgesellschaft, RoBdorf.

Slim, H. and P. Thompson, eds. 1993. Listening for a Change: Oral Testimony and Development. Panos Panos Oral Testimony Program. Panos Publications Ltd., London.

Subedi, A. and C. Garforth. 1996. Gender, Information and Communication Network: Implication for Extension. Eur. J. Agric. Educ. & Extension 32(2):63-74.

Triomphe, B. 1995. Agroecología del Sistema de Aboneras en el Litoral Atlántico de Honduras. Paper presented at the Manejo Productivo y Sostenible de las Laderas, XLI Reunión Anual del PCCMCA, Tegucigalpa, Honduras, 27 al 31 de marzo 1995.

Warren, D.M. and K. Cashman. 1988. Indigenous Knowledge for Sustainable Agriculture and Rural Development. Gatekeeper Series No. SA10: Briefing papers on key sustainability issues in agricultural development. International Institute for Environment and Development, London.

4. Integrating gender analysis for participatory genetic resources management: technical relevance, equity and impact

Maria Fernandez, Pratap Shrestha and Pablo Eyzaguirre

The need for gender analysis

Participatory approaches recognize and empower farmers and other stakeholders to participate in programmes for the management and sustainable use of plant genetic resources. The quality of participation, however, depends on how these stakeholders are identified, selected and represented in the process. Farming households differ greatly from each other in terms of access to and control over resources. There are also differences in the decision-making power that affect the allocation of resources and benefit accruing from research and development endeavours. Differences also exist among the members of a household, which influence the way farm households manage plant genetic resources in their community. We need to understand such differences and use the information in designing participatory approaches for the management of local plant genetic resources. Gender is the basis for fundamental differences in tasks, responsibilities and knowledge within households and communities in nearly all social groups. Thus gender analysis is an essential component of participatory approaches.

Conceptually, gender is a socially constructed (not sexually determined) way of being men and women. The roles of women and men change between societies, cultures and at different periods of time. According to Hannan-Andersson (1992), gender is a set of socioeconomic variables used to analyze the roles, responsibilities, constraints and potentials of women and men, and the social relations between them. Gender analysis, therefore, considers the roles of women and men in management, conservation and use of plant genetic resources within the context of intra- and inter-household dynamics in a given society.

Gender analysis combines a set of tools that enables us to understand gender differences and to use the knowledge to create more effective development processes through more equitable participation. It also enables us to focus on the various sets of knowledge, responsibilities and actions which affect the distribution and sustainable use of plant genetic resources. Evidence from previous research has indicated that the division of labour and knowledge between men and women is of great importance in determining crop uses and management.

While men's and women's responsibilities for the management of plant genetic resources vary greatly according to plant species, culture and household organization, women shape genetic diversity of crops and trees through their intimate knowledge of diverse uses and in their role in seed systems (Eyzaguirre and Raymond 1995). Studies in Asia, Africa and Latin America have shown that seed selection and storage is largely the domain of women, and that the women play a pivotal role in farmer-to-farmer distribution of seeds and other plant germplasm (Tapia and De la Torre 1998; Iriarte *et al.* 1999; Shrestha 1998). They are also important in the management of uncultivated crops for food and other household uses. Women are often responsible for securing medicines, craft materials and firewood from forest areas and their burdens may be substantially increased if their access to forestry resources is jeopardized (Wilde and Vainio-Mattila 1995). However, there are still constraints on the inclusion of women in participatory research and development programmes. Among these are the fact that women may not be included in the public domain or in community decision-making processes. Similarly, they may not identify with research questions, they may not be allowed to interact with male researchers, or they may be hindered by time constraints and inconvenience of mobility which do not allow them to participate in research and development activities. Applying gender analysis and participatory approaches should obviously rely on teams and processes in which both men and women are included, either separately or jointly depending on the local situation.

Given the central and multiple roles that women play, participatory methodologies for the management of plant genetic resources must include gender analysis in order to present an accurate picture of the social factors that shape the conservation and use of such resources. Gender analysis also provides a particularly clear example of where participatory methods can work to the benefit of both the researchers and participants. Gender analysis is also an important tool in demonstrating the targeted impact of PGR activities on agrarian communities.

Tools and practices for gender analysis

Gender analysis focuses on three sets of questions: who does what, when and where?
1. **gender activity analysis** documents the gendered pattern of activities and responsibilities within households and farming systems.
2. **gender access/control analysis** focuses on who has access to or control over resources for production.
3. **gender benefit analysis** focuses on who benefits from each enterprise.

Several manuals and examples of the use of gender-sensitive methods and tools (Box 4.1) in agricultural and rural development research are available (Feldstein and Jiggins 1994; CGIAR 1997; FAO 1998; IDRC 1998). In Feldstein and Jiggins (1994), various participatory research tools for gathering gender-specific information are presented along with case studies, which discuss their use in the field. IDRC (1998) focuses the use of such tools more specifically on the area of plant genetic resources management.

> **Box 4.1.** Tools for users and researchers
> - Resource maps and flow diagrams
> - Focus, interest and gender groups
> - Enterprise, labour use, seasonal calendars
> - Chronologies
> - Stakeholder linkage diagrams

Gender activity analysis

The gender activity analysis explores the range of activities for which women, men and children are responsible, and considers the ways such gendered activities are associated with the management of plant genetic resources. It specifically focuses on three questions: who does what, when and where? The other key questions to consider are:
- What is the time line of particular gender activities, i.e. are activities performed at specific times during a season, intermittently over the course of a season, on a daily basis?
- Do women/men/children engage in or rely upon social networks to perform key activities, such as plant collecting, seed sharing, plant processing?
- Are particular activities embedded within broader cultural practices and/or beliefs?
- How does the gender division of activities differ between men- and women-headed households?
- To what extent does the gender division of activities correspond to cultural perception of gender roles (e.g. women as mothers/child care providers, men as breadwinners)? In what ways do gender division of activities contradict cultural norms regarding gender roles and responsibilities? Do such contradictions suggest a process of social/cultural change?

In order to better understand the options and opportunities available to different producers engaged in management of plant genetic resources, it is crucial to examine the gender division of labour, i.e. what activities are women/men/children engaging in. The gender activity analysis reveals how the activities of women and men are inter-related and distinct. The allocation of responsibility for particular farm and off-farm tasks on the basis of gender, and the different form of knowledge that women and men hold regarding the properties and uses of plants, affect the status of genetic resources and distribution of genetic diversity (Table 4.1). The gender activity analysis of farm, household, community and market-oriented activities enables researchers to integrate gender-specific knowledge of biodiversity into research, to identify gender-specific interests and needs, and to target project-related assistance and support on that basis.

Gender access/control analysis

The gender access/control analysis examines who has access to and/or control over plant genetic resources, and how such access and control is exercised in the household and the community. IDRC (1998) has suggested the use of a simple tool called 'Mapping Gendered Spaces' for this purpose. The Gender Mapping is quite similar to the Resource Mapping (a PRA tool), in which access and use of different spaces, places and resources by women and men are specifically shown (Fig. 4.1). The gender access/control analysis explores and analyzes dominant sociocultural categories of 'women' and 'men', and multiple ways in which gendered uses of plant genetic resources conform to or contradict such expressions.

In the course of gender access/control analysis, the following key questions should be considered:
- What spaces, places and plant genetic resources are thought to be associated with the normative categories of 'women' and 'men'?
- What spaces, places and plant genetic resources are actually used by women and men? Do they conform to or contradict normative ideals associated with gender roles and gendered spaces?
- What is the significance of spaces, places and genetic resources to women and men?
- What spaces, places and plant genetic resources are jointly and/or independently controlled by women and men?
- Which spaces, places and plant genetic resources meet particular personal, practical and strategic gender needs?

Gender benefit analysis

The gender benefit analysis explores how women and men use different plant genetic resources for their benefits as well as for the benefits of the household and the community at large. The key questions to be considered are:
- What plant genetic resources are used by women and men, and what are they used for?
- Who collects and processes these plant resources for food and other uses, and how are these processed?
- Who holds the knowledge of processing of plant resources and the knowledge of using such preparations for different uses?
- What plant resources are commonly sold to local/regional markets, and by whom are they collected and sold?

Using the gender benefit analysis as a point of departure within interview sessions and focus group discussions allows for an in-depth examination and analysis of different kinds of knowledge pertaining to local plant genetic resources and their management. Such an examination reveals crucial information about who (i.e. women, men and children) is responsible for the collecting, processing, use and in some cases marketing of particular resource products/by-products, such as medicinal plants (Table 4.2). It also points out the kind of local knowledge that women and men and members of different ethnic groups possess with regard to collecting, processing and use of plant genetic resources.

Table 4.1. Gender activity analysis chart: activities related to biodiversity management

Activity	Gender/Age	Location	Time	Notations
Farming *(disaggregate by crop type)*				
land preparation	mostly men	farm		
nursery	women and girls	around home		
planting/ transplanting	women and children	home to farm		
weeding	women	farm		Many weeds have edible and/or medicinal qualities and are collected by women for home consumption/use
fertilizing	men	farm		
harvesting	mostly women	farm		
hauling	men and women	farm to home		
Food/plant processing and preparation				
harvesting	mostly women	farm		
collecting of edible and medicinal plants*	women and men**	roadside, forests		*Many plants have both food and medicinal uses in the home. ** Species dependent
processing of food crops and wild plants into edible form	women and women's groups	around home		
cooking	women and girls	around home		Many wild plants are edible only through special cooking procedures
processing and preparation of medicinal plants	local healers and women	home	year round	
administering medicines	local healers and women	home	year round	
Marketing				
sale of edible and medicinal plants to local/regional markets and buyers*	women and men**	local/regional markets	intermittent throughout rainy and parts of dry season	Buyers are more commonly women ** Women and men sell and control the income from different plant species
Seed selection and preservation				
on-farm selection	women and girls	farm/home	harvest	
seed drying*	women and girls	around home, rooftops**	post-harvest	* Sometimes involves cooking ** Seeds dried in a variety of places depending on seed type (shade/sunlight, hot/warm/cool)
seed storage	women	around home/ local seed bank	dry season	
seed exchange	village/market women	seed fairs, village nurseries, cultural events	dry season, year round	

Source: IDRC 1998.

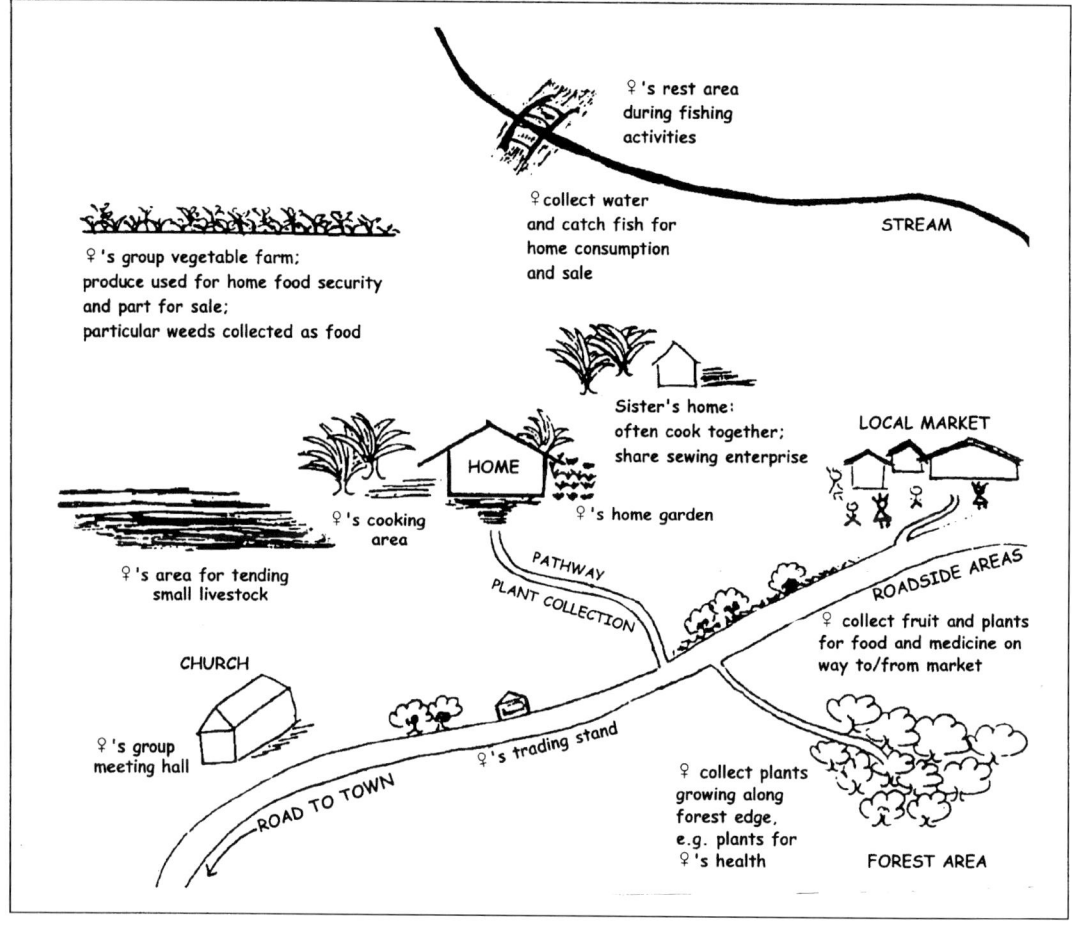

Fig. 4.1. Gender mapping (Source: IDRC 1998).

Practices of gender analysis

How gender analysis and tools are used in the field determines the quality of information gathered and the degree of participation of women and men in the process. Some of the following basic research practices are important for collecting gender-sensitive information:

- Interviews or exercises conducted separately for men's and women's groups: maps, transects, matrices, life histories, focus or community interviews, wealth ranking, venn diagramming, etc. Results of the separate exercises can then be compared to identify areas of both common and different knowledge and interests.
- Separate trials and field days to test technology options and discuss results.
- Researchers engage in participant observation in places where women work and with tasks done by women.
- Female researchers, field assistants and enumerators are included on the research team.
- In joint or separate meetings, questions are asked about tasks or enterprises which are known to be in the women's domain.
- Researchers collaborate with pre-existing women's groups.
- Researchers work with NGO partners that have access to women's groups.

Table 4.2. Benefits analysis chart: choice and use of local species by gender (sample)

Local name	Part used	Use(s)	Collection	Collected by	Location
(Plant A)	Leaves	Cooked in soups and stews (highly nutritious)	May-August	Women	Roadside pathways
(Plant B)	Leaves, twigs, seeds	Mat making/selling, oil	Aug-Nov	Women/girls Women/boys	Forest
Mushroom	Whole mushroom	Prepared in stews (by women)	March-Nov	Women	Forest floor
(Plant C)	Leaves	Used/sold as pain reliever (sold by men)	May-July	Men	Forest
(Plant D)	Leaves, bark	Processed and used by women for relief of menstrual discomfort and in large quantities may be used to induce miscarriage	April-July	Women	Grows wild among bean crop
(Plant E)	Leaves, buds	Used by men for hypertension	May-Sept.	Women/men	Forest
Tree (A)	Fruit, tree trunk	Prepared in meals/sold Used for building	June-Dec. Dec.-March	Women Men	Roadside

Source: IDRC 1998.

Gendered monitoring and evaluation

Often gender considerations in research and development programmes are limited to gender analysis. However, going beyond gender analysis is essential in order to assess whether the findings of gender analysis have been incorporated in the implementation of the programme and to examine the generated impact on women and men. This is where gendered monitoring and evaluation come into play.

Gendered monitoring tracks down how gender roles and needs, as revealed in the gender analysis, have been considered in the implementation of research and development activities related to the management of plant genetic resources. It also points out conditions/factors facilitating or obstructing the incorporation of gender issues in the implementation of a programme. Gendered evaluation, on the other hand, looks at the impact of consideration of gender issues on women and men in the management of plant genetic resources with the help of carefully chosen indicators.

Impact analysis and gender

Fortunately, today, impact analysis focuses increasingly on social actors and the extent to which development interventions, both technological and institutional, enhance their capacity to innovate and participate. In the words of Amartya Sen "Capacity reflects the liberty of a person to choose among different ways of life" (Sen 1989). For this reason, impact analysis should be geared to measuring increases in peoples' ability to choose from among diverse alternatives that can improve their quality of life (Box 4.2). If the alternatives are appropriate, access to them is equitable, and if people are empowered in the process, the impact will enhance the capacity of the individual, family and community to negotiate and to innovate.

Gender analysis allows us to understand the capacities and options available to people and how work on plant genetic resources has increased their capacity to sustainably manage and benefit from those resources. Using gendered participatory approaches will provide us with qualitative and quantitative impacts of PGR research and development processes. The process of identifying and selecting qualitative and quantitative indicators for measuring the impact of participatory PGR management on both men and women and other specific actors requires a selection of different tools. Both men and women must play an active part in the identification of the indicators to be used.

Characteristics of tools for assessing actor involvement:
- Focus on the development and resource management processes
- Characterize different kinds of participation
- Define limits of the situation where impact is to be assessed
- Develop appropriate impact criteria with technology users
- Organize information gathered for feedback to the community.

> **Box 4.2.** Measures of increased ability to assess and choose alternatives
> - Gender distribution of management responsibilities modified
>   - Patterns of control over resources
>   - Patterns of access to resources
>   - Changes in access to income
>   - Changes in use of income
> - Knowledge and skills improved:
>   - Increased knowledge of principles
>   - Enhanced skill in management practices
>   - Men's and women's knowledge is exchanged
> - Organization capacity enhanced:
>   - More/diverse stakeholders involved
>   - Leadership cadres increased
>   - Increased presence of women and men
>   - Women and men in leadership roles

Potential indicators for impact of participatory PGR management

The relevant variables for assessing the impact of participatory PGR management include a combination of physical access and the capacity of both women and men to envisage alternatives for their present and future use. This combination provides the basis for responsible decision-making and the innovation of management practices (Fernández and Salvatierra 1989).

Indicators that focus on the impacts of participatory PGR management on equity help to sort out how access and capacity to use genetic resources are affected by the power relations between men and women (Box 4.3). Where inequities exists, opportunities for increasing equity need to be realized through existing mechanisms and relationships within the community so that such changes are accepted and sustainable. Organizational indicators make it possible to measure the potential of a stakeholder group to negotiate with the larger society (Table 4.3). This is particularly important for *in situ* conservation of plant genetic resources.

> **Box 4.3.** Who has what and how much?
> - Access to and control over PGR
> - Technical skills and knowledge
> - Space for responsibility and decision-making
> - Information and influence
> - Income and capacity to invest

The capacity to innovate is indirectly related to an understanding of physical and organizational processes underpinning resource management strategies and the ability to translate visions of the future into present action. The first step in increasing access, participation and decision-making is to understand relations of power and communication among groups of actors. Impact studies encourage us to clarify our view of what is going on and with whom. They encourage us not only to focus on tools, but also to be very clear on what our indicators are. The issues and tools that follow can help us measure the impact of participatory approaches to plant genetic resources management on the capacity of users and stakeholders to innovate and derive benefits (Ortiz, unpublished).

Table 4.3. Where gender counts most

	Gender	Wealth	Generation	Ethnicity
Individual	++++	+++	+	+
Family	++++	+++	+++	+
Community	++++	++++	+++	+
Region	++	++	+++	+++
Nation	+	+	++++	++++

Conclusion

The integration of gender analysis is a prerequisite for participatory methods used in the management of local plant genetic resources. It enhances the quality of participation and the participation of the relevant and targeted users and beneficiaries of plant genetic resources. When applying these tools in practice one should ensure that gender analysis is not used in a mechanical way but is used to produce desired gender impacts. For this reason, the application of gender analysis tools should be linked to the measurement of impacts and benefits from the implementation of participatory PGR management processes. The use of tools for gender analysis and gendered monitoring and evaluation need to be continually examined, selected and modified if required to suit the local conditions. Local communities and PGR users can usefully participate in this process as well.

Acknowledgements

We acknowledge the IDRC-funded 'Guidelines for Integrating Gender Analysis into Biodiversity Research', Sustainable Use of Biodiversity (1998), and the System-wide Program on Participatory Research and Gender Analysis of the CGIAR for the contribution of the tools and methods included in this article.

References

CGIAR, SWP/PRGA. 1997. New Frontiers in Participatory Research and Gender Analysis. Proceedings of the International Seminar on Participatory Research and Gender Analysis for Technology Development, 9-14 September 1996, Cali. CGIAR.

Eyzaguirre, P.B. and Ruth Raymond. 1995. Rural Women: A Key to the Conservation and Sustainable Use of Agricultural Biodiversity. Fourth World Conference on Women- Focus Day on Rural Women, FAO, Beijing

FAO. 1998. Socioeconomic and Gender Analysis (SEAGA): Field Handbook. Food and Agriculture Organization of the United Nations, Rome.

Feldstein, H. and J. Jiggins (eds.) 1994. Tools for the Field: Methodologies Handbook for Gender Analysis in Agriculture. Kumarian Press, Connecticut.

Fernández, M. and H. Salvatierra. 1989. Participatory Technology Validation in Highland Communities of Peru. Pp. 146-150 *in* Farmer First: farmer innovation and agricultural research (R. Chambers, A. Pacey and L. Thrupp, eds.). Intermediate Technology Publications, London.

Hannan-Andersson, C. 1992. Gender Planning Methodology: Three Papers on Incorporating the Gender Approach in Development Cooperation Programmes. Rapporteur Och Notiser 109. Institution for Kuffurgeografi och ekonomik geografi unid lund Universitet.

IDRC. 1998. Guidelines for Integrating Gender Analysis into Biodiversity Research. Sustainable Use of Biodiversity Program Initiative, IDRC, Ottawa.

Iriarte, Lucio, Litza Lazarte, Javier Franco y David Fernández. 1999. El Rol del Genero en la Conservación, Localización, y Manejo de la Diversidad Genética de Papa, Tarwi y Maiz. Gender and Genetic Resources Management. IPGRI, FAO, BIOSOMA, Rome.

Sen, A. 1989. Development as capability expansion. J. Development Planning No. 19.

Shrestha, P.K. 1998. Gene, Gender and Generation: Role of Traditional Seed Supply Systems in the Maintenance of Agrobiodiversity in Nepal. Pp. 143-152 *in* Managing Agrobiodiversity: Farmers' changing perspectives and institutional responses in the Hindu Kush-Himalayan region (Tej Partap and B. Sthapit, eds.). ICIMOD, IPGRI, Kathmandu.

Tapia, Mario E. and Ana De la Torre. 1998. Women Farmers and Andean Seeds. Gender and Genetic Resources Management. IPGRI/FAO, Rome.

Wilde, V. and A. Vainio-Mattila. 1995. Gender Analysis and Forestry: International Training Package. FAO, Rome.

Section II

Enhancing farmers' access to plant genetic resources maintained *ex situ*

5. Overview

Esbern Friis-Hansen and Bhuwon Sthapit

Preservation of genetic diversity is the principal concern of centralized *ex situ* genebanks and plant breeders, the main users of genebanks. *Ex situ* genebanks have, until recently, rarely used participatory approaches to identify, collect, characterize, document, store and maintain genetic diversity. Methods for collecting genetically diverse species have been developed by conservation scientists with the aim of best capturing the widest genetic diversity available at a given location within the relatively strict time limits usually applied to plant genetic resources collecting missions. Such biologically defined methods exclude farmer's active participation and commonly reduce their roles to that of passive providers of planting materials.

The non-participatory approach of traditional genebanks has not been criticized by conventional plant breeders as their principal interest is access to the widest possible selection of genetic material without concern for the participatory methodological aspects of its collection. The assumptions behind the traditional top-down approach have therefore remained largely unchallenged. However, both from a scientific perspective and from a developmental perspective, there are convincing arguments suggesting the potential benefits to genebanks of their engaging in participatory approaches and establishing a dialogue with farmers and their communities. NGOs have been voicing the view that genebanks should develop ways and means to deploy diversity back to the farmers' fields.

Participatory approaches have been suggested and, in certain cases, tested, within the genebank activities of collecting of plant genetic resources and compiling of knowledge related to stored samples (i.e. passport data). Participatory germplasm collecting uses local farmers, professional NGOs and state employees to plan and conduct plant collecting. Local collectors often have extensive knowledge of ecogeographical, biological and cultural issues, are present year round and can collect throughout the fruiting season (which may be important for crops with large variations in time to maturity) and are better placed to include important microenvironments within complex agricultural systems which may otherwise be overlooked. Developing participatory relationships with local partners may furthermore be cost-efficient, as shortages of funds, human resources and equipment often severely restrict the ability of national genebanks to initiate collecting missions (Nazarea-Sandoval 1990; Guarino and Friis-Hansen 1995). In recent years, IPGRI's Global Project on the Scientific Basis of *In situ* Crop Conservation in Vietnam and Nepal has used biodiversity fairs as a means to locate useful diversity. This activity also has been found to be an effective participatory method to collect germplasm as well as associated indigenous knowledge.

New and potential users of genebanks are emerging. These users are likely to require and demand a different type of service and product from the genebank. These are, on the one hand, NGOs, community-based organizations, relief or on-farm conservation programmes and other donor-financed projects which seek to enhance or reintroduce species/varietal diversity in farming communities. Other users include scientists involved in participatory variety selection and participatory plant breeding. Such users of genebanks will be more likely to demand farmer's original landraces/varieties than segregated lines. They also will require passport information relating to these farmers' varieties, including indigenous knowledge about their useful traits and end-uses.

Some National Genebanks are still reluctant to broaden the base of users and be more proactive in diversity deployment strategies. However, this trend is changing and some new initiatives have been taken by national programmes.

In 1988 the National Genebank of Ethiopia pioneered by directly involving farmers in its plant genetic resources activities. Chapter 6 reviews the Ethiopian experience and is written by the key persons involved with transforming that genebank's approach. Chapter 7 discusses the experience of the plant genetic resources programme at Can Tho University in Vietnam. As part of the CBDC programme, Can Tho University has gained experience with linking its rice genebank with farmers through PPB and PVS. The success of this programme is replicated elsewhere in Vietnam.

Chapter 8 provides a historic view of different collecting strategies and goals at the ICRISAT genebank. The chapter first provides an account of the conventional collecting and conservation strategies which are organized by and for plant breeders. The chapter thereafter outlines the rethinking in the second half of the 1990s within the ICRISAT genebank and its consequences for changes in collecting strategies, documenting farmers' knowledge, linking farmers through PPB and enhancing farmers' access to germplasm.

The potential role of genebanks as a source of diversity deployment, in post-war times or following drought or other natural calamities, is discussed in Chapter 9, using the case of Somalia. Unlike other emergency seed-supply programmes, this programme attempted a participatory approach including a socioeconomic and farming system baseline survey, on-farm trials and PVS of re-introduced germplasm.

Future challenges for genebanks are that they not only collect and catalogue genetic resources from the farming community but also use participatory approaches to deploy genetic resources to support the productive activities of farmers. Strengthening grassroots seed networks, diversity fairs and exchange and participatory plant breeding are some options which may add benefits to local diversity and provide benefits to local communities. Farmers participating in PPB programmes benefit from early access to new genetic materials and contribute to the improvement of their livelihoods.

References

Guarino, L. and E. Friis-Hansen. 1995. Collecting plant genetic resources and documenting associated indigenous knowledge in the field: a participatory approach. Pp. 345-361 *in* Collecting Plant Genetic Diversity (L. Guarino, V.R. Rao and R. Reid, eds.). CAB International, Oxon, UK.

Nazarea-Sandoval, V.N. 1990. Potentials and limitations of ethnoscientific methods in agricultural reserarch. *In* Incountry Training Workshop for Farm Household Diagnostic Skills (R.E. Rhodes, V.N. Sandoval and C.P. Bagalanon, eds.). CIP, Los Banos.

6. Participatory approaches linking farmer access to genebanks: Ethiopia

Melaku Worede, Awegechew Teshome and Tesfaye Tesemma

Introduction

Ethiopia is one of the eight centres in the world where crop plant diversity is strikingly high and is a centre where some crop species were domesticated (Vavilov 1926, 1951). Ethiopia is also a region where the traditional farming systems have co-evolved with the diverse Farmers' Varieties (FVs) over millennia. Over the last 25 years Ethiopia's diverse genetic resources, however, have faced threats of genetic vulnerability due to severe recurrent droughts that also afflicted other parts of Africa and risk of displacement of the diverse FVs by high-yielding commercial varieties (HYVs). Where the HYVs were planted, the land-use systems changed to meet their demands, with resulting habitat destruction in both the wild and managed ecosystems of the country (Worede 1992). Where FVs are lost, the traditional knowledge of cropping patterns and management practices and the ecological rationale behind them are also lost.

An important step in meeting the challenges of food security, agricultural sustainability and maintaining genetic diversity is to forge a partnership between the formal and informal systems. One way of fostering such a partnership is to encourage farmers and scientists to work together in participatory genetic resources activities by incorporating time-tested farmers' knowledge in the sustainable management of genetic resources, especially in marginal and heterogeneous environments.

This paper highlights the partnership forged between farmers and scientists from the national genebank and various collaborating institutions in the rescue, collecting, multiplication, maintenance, enhancement and utilization of existing diverse crop genetic resources in Ethiopia. The paper also describes on-farm genetic resources activities that assist farmers in keeping their diversity while raising productivity, and influencing policy pertaining to such an exercise.

The Ethiopian Genebank

The concern over the alarming rate at which genetic erosion is progressing in Ethiopia stimulated interest among national and international scientists that resulted in the establishment of the Plant Genetic Resources Center (PGRC/Ethiopia) as a national genebank in 1976. The Ethiopian genebank at present holds more than 56 000 accessions of various crop plant species as long- and medium-term collections.

Genetic resources activity has been undertaken systematically over the last 22 years and represents a major national effort of the Ethiopian Plant Genetic Resources Center (PGRC/Ethiopia), now the Institute of Biodiversity Conservation and Research (IBCR). The genebank salvages and utilizes farmers' varieties through a complementary approach involving farmer-based activities and *ex situ* (off-farm) conservation in genebanks.

Since its inception, the genebank has played an exemplary role in undertaking collaborative *ex situ* conservation activities with other international genebanks around the world. However, it was realized in due course that the *ex situ* measures alone would be inadequate in protecting the country's crop genetic resources, the bulk of which are still managed by farmers who represent 85% of Ethiopia's population. The livelihood security of the Ethiopian farmers is dependent on the sustainable use of the diverse crop genetic resources they maintain in the fields. Complementing the *ex situ* (off-farm) conservation with a more dynamic conservation measure was, therefore, considered crucial to sustaining diversity. This led to the idea of *in situ* (on-site) conservation of farmers' varieties on farmers' fields. *In situ* conservation is an evolutionary process in which the biotic and abiotic factors of an agro-ecosystem dynamically interact with farmers' selection criteria, resulting in diverse genetic resources to meet the multiple needs of farmers and adaptation requirements of the agricultural habitats in which the crops are grown. Thus, such a conservation approach is essentially linked to rural development, farmers' ethnobotanical knowledge, farming practices, selection criteria and innovations.

Farmers: key players in sustaining diversity

Farmers are creators, managers and primary users of the biological diversity generated on their fields. They have extensive knowledge of their agricultural production systems. Farmers employ multiple strategies to generate and maintain genetic diversity. The generation and maintenance of the immense diversity that we see in the field is not a random occurrence. The farmers play a substantial role by applying their understanding of the elements and interactions of the agro-ecosystem, guided by a relatively sophisticated folk taxonomic classification, farming practices and selection criteria (Teshome 1996; Teshome *et al.* 1997, 1999b). One aspect of traditional storability knowledge was remarkably precise in predicting observed insect resistance (Teshome 1996; Teshome *et al.* 1999a). Farmers' knowledge of storability is used to reduce the risk of loss of a major food supply as well as genetic diversity due to storage pest infestations. These and other dynamic and adaptable farmers' knowledge adds to the knowledge of the scientific community, particularly as to how farmers generate, select and maintain diversity in their fields.

In Ethiopia, traditional farming represents centuries of accumulated experience and skills of farmers who have farmed under diverse conditions using locally available resources that often resulted in stable yields. The basis for this is the traditional crops and their landraces which farmers have adapted over centuries of selection and use to meet dynamic and changing needs. Ethiopian farmers are also instrumental in conserving germplasm as they control the bulk of the country's genetic resources. Farmers always retain some seed stock for security unless circumstances dictate otherwise. In various regions of the country, farmers also have well-established systems of ensuring sustenance of seed supply and they often operate in networks. These include exchange of seed and knowledge of crop types having a wide range of adaptation to diverse environments.

Establishing the farmer-genebank link

Ethiopian farmers and the national genebank began their collaboration in the late 1970s when the genebank and its international partners embarked on extensive genetic resources rescue and collecting missions in Ethiopia. Germplasm collections were based on the dynamic and time-tested ethnobotanical knowledge of local farmers used for centuries to characterize, select and utilize their crop genetic resources *in situ*.

In the late 1980s, in an effort to mitigate the impact of drought on the country's genetic resources, the Ethiopian genebank in collaboration with farmers and financial support provided by a Canadian NGO, the Unitarian Services Committee of Canada (USC/C), embarked on rescue, conservation, multiplication and enhancing the performance of the invaluable heritage of traditional farmers' seeds on farmers' fields. The participatory programme was initiated to undertake farmer-based conservation and utilization of plant genetic diversity in stress-prone areas, with a view to promoting the food and livelihood security of small-scale farmers in the country. This task was officially delegated by the national genebank to the USC/C–supported Seeds of Survival (SoS/Ethiopia) programme which carried out the work diligently, professionally and successfully until September 1997 when it handed over all the ongoing genetic resources activities to the Ethiopian Institute of Biodiversity (BDI).

Two major approaches were considered in undertaking such participatory genetic resources activities under which farmers and the genebank scientists were collaborating in the rescue, collecting, conservation, maintenance, enhancement and utilization of diverse crop genetic resources in the central highlands of Ethiopia (Worede and Mekbib 1993). The first approach was the rescue, conservation and restoration of FVs on-farm. Second was the maintenance of elite landrace selections on-farm.

The overall goal of the participatory activities was to raise productivity while maintaining an appreciable amount of diversity of seed materials at the community level. Farmers' selection criteria represent a crucial input in ensuring that the retention of diversity and productivity characters were included in formulating the overall goal of the programme. It was considered crucial to tap existing local knowledge by establishing measures to conserve *in situ* the crop landraces considered in the participatory programme.

The programme primarily represented a participatory, dynamic, farmer-based approach to conservation, enhancement and utilization of FVs. Moreover, activities were linked to the more formal off-farm conservation activities of the national genebank. The work had been carried out on small-scale farmers' fields with full participation of farmers, scientists and local extension agents.

Rescue, conservation and restoration of farmers' varieties on-farm

This aspect of participatory genetic resources activity involves farmers, scientists and extension workers and has been in progress since 1988, initially embracing 500 farmers selected and organized through their respective cooperatives in north Shewa and Welo regions of Ethiopia. The project sites support a rich diversity of crop genetic resources and are susceptible to recurrent drought and other stresses, including disease and pest epidemics. The crops include sorghum, chickpea, field peas and locally adapted maize (Fig. 6.1).

This participatory activity was designed primarily to maintain crop diversity in a dynamic state on farmers' fields with a view to protecting major cultivars from disappearance by improving their genetic performance through conservation, enhancement and use practices. Sorghum, for its varied uses in the community and its resilience to environmental stochasticity in the agro-ecosystem, was selected by the participating farmers and scientists as a prime research crop in the participatory programme.

Sorghum seed materials were obtained from the participating farmers and *ex situ* collections of the national genebank. The seed materials were distributed to the participating farmers representing diverse and heterogeneous agricultural environments. They were encouraged to grow them on their respective fields following time-tested farming practices of selection, thinning, weeding, harvesting, production, storage and uses. Crops grown on each field vary seasonally, according to the individual farmer's use of traditional cropping patterns and associations of various inter- and intra-specific crops grown in the respective fields of the project area.

Each participating farmer determined the seeding rate based on experiences of need, amount of seed and labour availability, method of seeding, and soil and climatic variations. In the participatory project, sorghum was planted in plots ranging from 0.5 to 4 ha with a seeding rate of 5 to 20 kg/ha. Farmers' rationales for the minimum plot size and seeding rate requirements are to ensure a safe harvest and maintain crop genetic diversity as has been done for centuries and thus provide the basis for a sound and viable conservation approach. Partner farmers practised various forms of mass and stratified selections, and multiplied their crop landraces, mainly sorghum and maize, separately for production. Seeds of selected sorghum plants were bulked to form a slightly improved population to be used as sources of seed production and continued selection. This was done primarily on the basis of their own traditionally established and diverse selection criteria complemented by a descriptor list provided by the national genebank and SoS/Ethiopia. Farmers' yield criteria demonstrated that a significant improvement in crop yield has been observed, by participating farmers and scientists, among the selected materials grown under farmers' conditions *in situ*.

Fig. 6.1. Landscape in N. Shewa (Photo courtesy of Awegechew Teshome).

Before the actual harvest, participating farmers and scientists from the genebank walk around inside each sorghum field and select representative sorghum heads to be used by the farmers as sources of seed for the next planting season and as germplasm samples for storage, further evaluation and characterization at the national genebank. The evaluated and characterized germplasm will be used in local and institutional crop improvement programmes as sources of genes that control characters of interest including disease/pest resistance, essential amino acid content (such as high lysine) and drought tolerance. The selections will conserve and enhance potential elite intra-specific FVs *in situ*.

Close to 500 farmers actively participated and benefited from the seed developed by the programme and multiplied the seed on a continuous basis for further distribution to other farmers of the region. Most of the sorghum landraces and locally adapted maize jointly selected with farmers did very well with no external inputs of fertilizers and pesticides. These materials have expanded into the larger parts of the project regions where frequent crop failures have occurred owing to prevailing droughts, thus filling major gaps in the availability of locally adapted seeds for planting under such stressful conditions. This participatory activity has also helped increase the amount of diversity now existing in the project region and adjacent areas as a result of the restoration and crop-enrichment efforts by both farmers and scientists. An independent study (Teshome 1996) has demonstrated that there are 60 agromorphologically distinct intra-specific sorghum FVs representing the crop genetic diversity richness of sorghum in the project area. This intra-specific sorghum richness is the result of the relatively sophisticated folk taxonomy (Teshome *et al.* 1997), farming practices and selection criteria (Teshome *et al.* 1999b) employed by the participating farmers and their communities in the agricultural landscape of the project area.

The participatory efforts of farmers and scientists have saved from extinction several cultivars of sorghum, wheat, maize and the associated crop species that were maintained, conserved and utilized for centuries *in situ* by farmers.

Maintenance of elite durum wheat landrace selections on-farm

This participatory activity involved farmers and scientists in promoting the conservation, enhancement and utilization of indigenous durum wheat in Adaa district of central Shewa region in the central highlands of Ethiopia. The main reason for the participatory programme was that the indigenous durum wheat was almost disappearing because of the threat of displacement by introduced bread and durum wheat varieties. In the project area women farmers traditionally use the local durum to make porridge, *enjera* (unleavened bread), home-brewed beer and sell and/or exchange it at the local markets. They rarely use bread wheat for themselves; rather, they sell it as a commodity crop in the urban markets.

The participatory materials were collected, selected and upgraded on a continual basis to improve their competitiveness. These materials were obtained from the national genebank and the Debre Zeit Agricultural Research Center. Fifty composites were selected for multiplication and evaluation by participating farmers. Elite materials were selected and developed, based on yield performance tests under different stress conditions, by the Debre Zeit Agricultural Research Center (Tesemma 1987) and were handed out to farmers for further selection on-farm by collaborating farmers and scientists. The eight composites most preferred by the farmers have been in production at various agro-ecological sites represented by 4000 farmers' fields ranging in size from 0.5 to 10 ha. Some of the elite durum wheat selections (composites of 2-5 agromorphotypes) developed by this participatory approach were assessed by the farmers as being more productive. Farmers' elite varieties grown without the use of commercial fertilizers or other chemicals outyielded their high-input counterparts represented by durum wheat HYVs, on average by as much as 10% (Ataro and Bayush 1994).

The farmers continue to multiply and use the composites that are best suited to their conditions along with other entries provided by the participating breeder scientists. The national genebank takes representative samples from these lines for long-term storage. This allows the farmers to critically evaluate the sources, adaptability and performance of planting materials of their farming systems by comparing their own materials with off-farm sources. It also encourages them to make continued use of their locally adapted intra-specific farmers' varieties, to ensure effective utilization of superior germplasm, by avoiding the threat of losing unexplored variation within and between the indigenous populations. Besides, a number of selected farmers are multiplying the elite seeds for distribution to the surrounding farms through market and other cultural and social diffusion mechanisms. Currently, at Ejere in central Shewa Region, a local community maintains 100 quintals (10 000 kg) of seed stock (mainly durum wheat composites) at its Community Seed Bank (CSB) built by IBCR with some 500 local farmers having access to such materials (Regassa Feyissa, pers. comm.). Representative samples also are maintained *ex situ* at the genebank.

Although the focus is on elite materials, the original seed stock is included in such a strategy and maintained *in situ* with the traditional methods and practices employed by the participating farmers of the project area. Other crops, particularly pulses (chickpea, faba bean and fenugreek) and barley, are grown in rotation with elite materials.

Linking conservation to utilization

The value of farmers' varieties to farmers in a developing country like Ethiopia lies in their utility as a dependable source of planting and breeding materials. It is, therefore, important that locally adapted and enhanced seed materials are multiplied for distribution to farmers whose requirements have not been adequately met by modern high-input cultivars. The enhanced materials are potentially marketable to improve the income and livelihood of the farmers. A seed-supply system is currently being developed by the Ethiopian genebank involving different stakeholders participating in the conservation, enhancement and utilization of crop genetic resources in Ethiopia.

The complementarity of the *ex situ* and *in situ* conservation approaches is clearly indicated through monitoring, restoration, introduction, research and extension activities. Representatives of what has been generated and maintained by farmers on-farm are selected, multiplied, stored in community genebanks, reach the local markets and are distributed through extension, traditional and social diffusion mechanisms in the community and beyond its agricultural landscape. Representative germplasm materials stored in the national genebank are brought into farmers' fields and community use after passing through the scientific selection, multiplication, breeding and production processes of the formal systems of research, crop improvement and extension services. The two systems meet through their genetic materials in the community and on farmers' fields, further generating and diversifying crop genetic resources of various types at the different levels of the agro-ecosystem. The *in situ* approach brings dynamically generated new material, as a result of the spatial and temporal interactions of farmers' uses and selections, with the biotic and abiotic elements of the agroecosystem. The *ex situ* approach plays primarily a back-up role for any unforeseen losses that may happen in the field. Thus, we need to have the complementarity of the two conservation approaches to meet the current challenges of agricultural sustainability, food security and agrobiodiversity conservation by forging a dynamic partnership between farmers and scientists in participatory genetic resources activities.

References

Ataro Adare and Bayush Tsegaye. 1994. Survey on relative performance of landraces for the 1993/94 cropping season. Consultancy Report, Seeds of Survival, USC/Canada in Ethiopia.

Teshome, A., J.T. Arnason, J.K. Torrance, J.D. Lambert, L. Fahrig and B. Baum. 1999a. Traditional farmers' knowledge of sorghum landrace storability in Ethiopia. Econ. Bot. 53(1):69-78.

Teshome, A., B. Baum, L. Fahrig, J.K. Torrance, J.T. Arnason and J.D. Lambert. 1997. Sorghum landrace variation and classification in north Shewa and south, Ethiopia. Euphytica 97:255-263.

Teshome, A., L. Fahrig, J.K. Torrance, T.J. Arnason, J.D. Lambert and B. Baum. 1999b. Maintenance of sorghum landrace diversity by farmers' selection in Ethiopia. Econ. Bot. 53(1):79-88.

Teshome, Awegechew. 1996. Factors Maintaining Sorghum Landrace Diversity in north Shewa and south Welo regions of Ethiopia. PhD Thesis. Ottawa-Carleton Institute of Biology, Biology Department, Ottawa, Canada.

Tessema, Tesfaye. 1987. Improvement of indigenous wheat landraces in Ethiopia. Pp. 232-238 *in* Proceedings of the International Symposium on Conservation and Utilization of Ethiopian Germplasm, 13 –16 Oct. 1986 (J.M.M. Engels, ed.).

Vavilov, N.I. 1926. Studies on the origin of cultivated plants. Russian and English, State press, Leningrad. 248.

Vavilov, N.I. 1951. The origin, variation, immunity, and breeding of cultivated plants. *In* Selected writings of N.I. Vavilov. Translation by K. Starr, Chester. Chronica Botanica 13(1/16).

Worede, M. and H. Mekbib. 1993. Linking genetic resources conservation to farmers in Ethiopia. Pp. 78-84 *in* Cultivating Knowledge: Genetic diversity, farmer experimentation and crop research (Walter de Boef *et al.*, eds.). Intermediate Technology Publications, London.

Worede, Melaku. 1992. Ethiopia: A gene bank working with farmers. Pp. 78-94. *in* Growing Diversity (D. Cooper, Renee Vellve and Henk Hobbelink, eds.). Intermediate Technology Publications, London.

7. Linking the national genebank of Vietnam and farmers

Nguyen Ngoc De

Introduction

In Vietnam, plant genetic resources (PGR) conservation and utilization have drawn much attention from many scientists and the governments, especially following the war. Khoi (1996) reported that Vietnam owns about 700 crop species belonging to 70 genera. At present, there are about 39 species of starchy food crops, 95 species of non-starchy food crops, 104 species of fruit crops and about 55 species of vegetables.

Genetic resources of major crops – especially rice, root crops and beans – have been collected, evaluated and conserved by specific research institutions. The national genebank operated by Vietnam Agricultural Science and Technology Institute (Hanoi) has the largest collection of PGRs of major crops from across the country. In addition, other specific-crop research institutes have their own collections and genebanks for their own research and breeding programmes. Germplasm exchange occasionally has taken place among research institutions and under request from breeders.

In the Mekong Delta of Vietnam, Can Tho University has its own collections of rice and beans. Its collecting activities have been carried out since 1994. At present, the University has 1552 accessions of local traditional rice from the Mekong Delta, 491 of upland rice from the western highlands and northern mountainous areas, and 10 of wild rice species. Characterization and evaluation have been done and valuable accessions for stress tolerance, pest resistance and grain quality have been used in breeding programmes. The operation of the University's genebank differs from that of the other genebanks.

Farmers have played a very important role in the whole process of diversifying plant and animal genetic resources. They have utilized, domesticated, conserved, improved and developed PGRs to meet their needs, and in the process have maintained and conserved for future generations. However, most of the modern breeding activities and the introduction of new crop varieties have been done by formal sector plant breeders. Linkage between breeders and farmers is generally weak. The role of farmers in the crop-improvement process has not been well appreciated. Farmers are, therefore, treated as passive users of the formal-sector plant breeding products. As a result, the adaptation of the recommended varieties, in many cases, is very slow, doubtful and has even failed.

In recent years, participatory approaches have been applied in many rural development programmes in Vietnam. Can Tho University, as the leading research institution in adopting a participatory approach in rice improvement, started on-farm breeding programmes as early as 1975 (after the war) by sending out their staff and students to work closely with farmers on problem identification, priority-setting, testing of selected breeding materials and reintroduction of collected landraces. Genebank accessions and breeding materials have been tested for tolerance of acid sulphate soils, deepwater growing conditions and salinity in the real soil conditions. In 1994, with the inception of the Community Biodiversity Development and Conservation (CBDC) project, Participatory Plant Breeding (PPB) and Participatory Variety Selection (PVS) were introduced to develop and identify crop varieties specific to the niche environments and farmers' preferences (Fig. 7.1). PPB and PVS have helped to assist, support, strengthen and enhance farming communities in improving and diversifying their local plant genetic resources by bringing breeders closer to farmers and by strengthening the farmers' role in plant breeding and management of local PGRs. Farmers – under their natural, socioeconomic and political environments – have their own needs, purposes and criteria for crop improvement. PPB and PVS have opened up ways to empower farmers and farming communities in the decision-making processes on their choices and improvement of crops and crop varieties.

Reasons for linking genebank and farmers

The most important reasons for linking the genebank to farmers and farming communities is based on the mutual benefits between breeders and farmers, research institutions and farming communities. Breeders need to test their breeding materials to fulfil their research objectives. Farmers need to have more choices in crop improvement for their particular environments and needs. This is the most cost-effective research approach. Farmers provide land for testing and partial labour for crop care. It was meaningful for research institutions in the transitional period after the war when funding for research was very limited but demand for rice improvement was very high. Can Tho University (Rice Research Department) initiated its on-farm testing/trials, and PPB/PVS programmes in the Mekong Delta in this environment. Linking was, therefore, a primary objective of the work undertaken.

The institutions engage in PPB/PVS (Participatory Varietal Selection) including farmers/communities, breeders/scientists, research institutions, seed production centres/companies and agricultural extension centres. The key players from the formal sector are breeders, scientists, researchers, technicians and extensionists; from the informal sector, the key players are farmers and farmer communities.

The linkage between genebank and farmers is two-way: breeders and research institutions get access to actual conditions for testing their breeding materials; farmers and farming communities can ask for specific breeding materials from the genebank and research institutions. Figure 7.2 shows these linkages, the key players and the contribution of each sector.

The formal-sector roles are breeding, selecting, testing and providing breeding materials/seeds for farmers, assisting/training/transferring advanced technologies, and getting feedback from farmers. The informal sector can also be involved in breeding, selecting, testing, evaluating, multiplication, distributing seeds, and reflecting results, problems and requests to the formal sector. A successful PPB programme is one that involves farmers and farmer communities as much as possible in the whole process of plant breeding. They are partners in term of collaborating/linking activities in PPB/PVS, sharing/exchanging seeds/breeding materials and working together.

Processes in linking with genebank

Every year Can Tho University provides hundreds of testing varieties to farmers under their testing programmes. Farmers and farming communities can also make their requests for particular varieties or breeding materials to the genebank for their own purposes. However, the linking of farmers with genebank is mainly through PPB and PVS. This is the common way to empower farmers and farming communities in crop improvement (Fig. 7.3).

The steps/processes/activities undertaken that facilitated linking national genebank and farmers are as follows.

Site selection

Criteria for site selection:
- The project objectives should fit the farmers' needs
- Local conditions including farming systems, cultural practices and socioeconomic conditions should fulfill the project requirement
- Farmer's needs and willingness to collaborate are very important factors for implementation
- Support from local organizations is needed to facilitate the project activities.

Methods used for site selection:
- Agro-ecosystem characterization and analysis from survey and mapping techniques are used to describe the sites
- Community meetings, personal communication with local extension workers and experienced farmers are arranged to get general information on related issues.

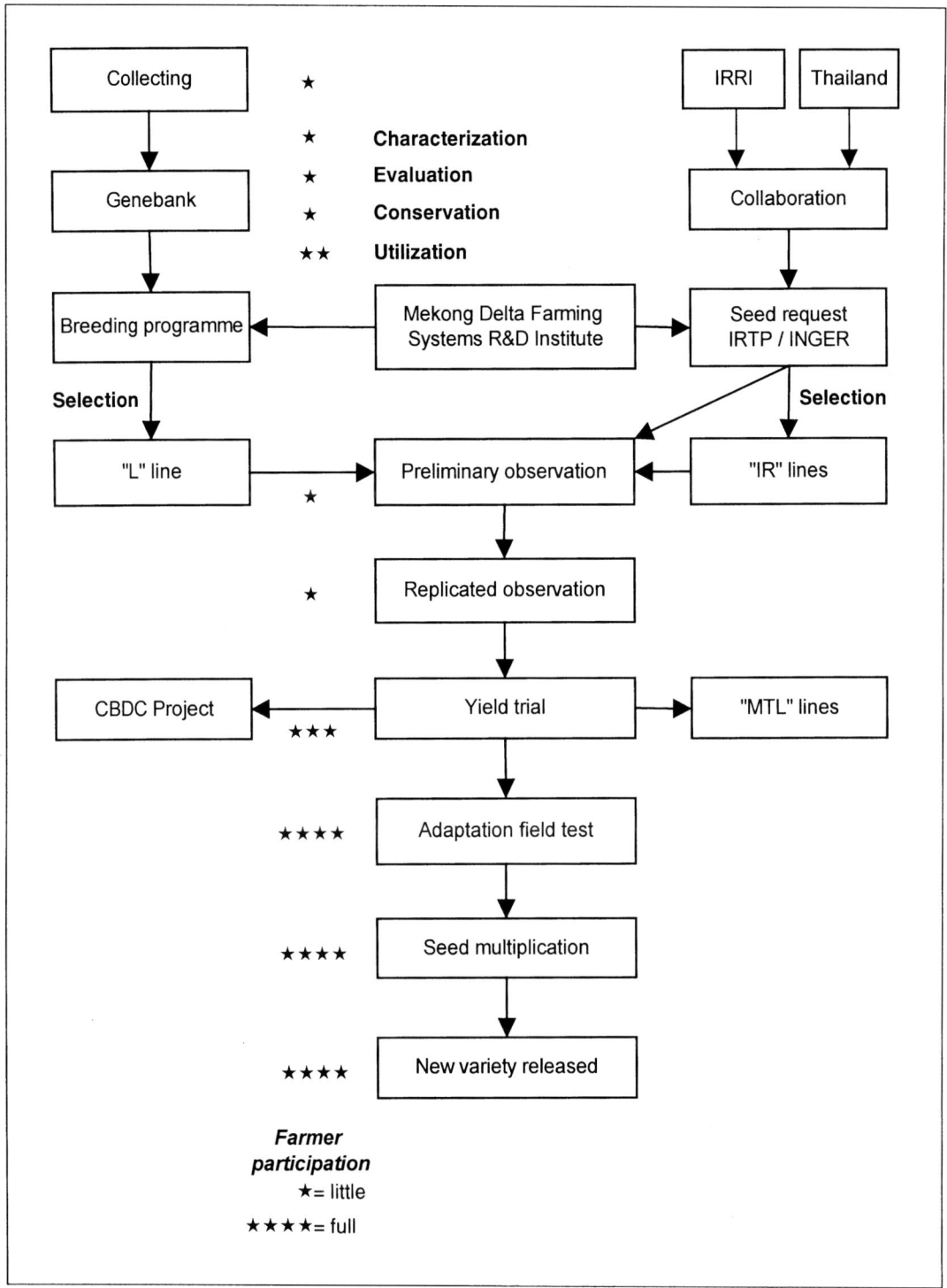

Fig. 7.1. Participatory Rice Varietal Improvement Programme in Farming Systems Research and Development Institute, Can Tho University, Vietnam. Note: MTL (Mien Tay Lua) was crossed, selected and released by Farming Systems Institute.

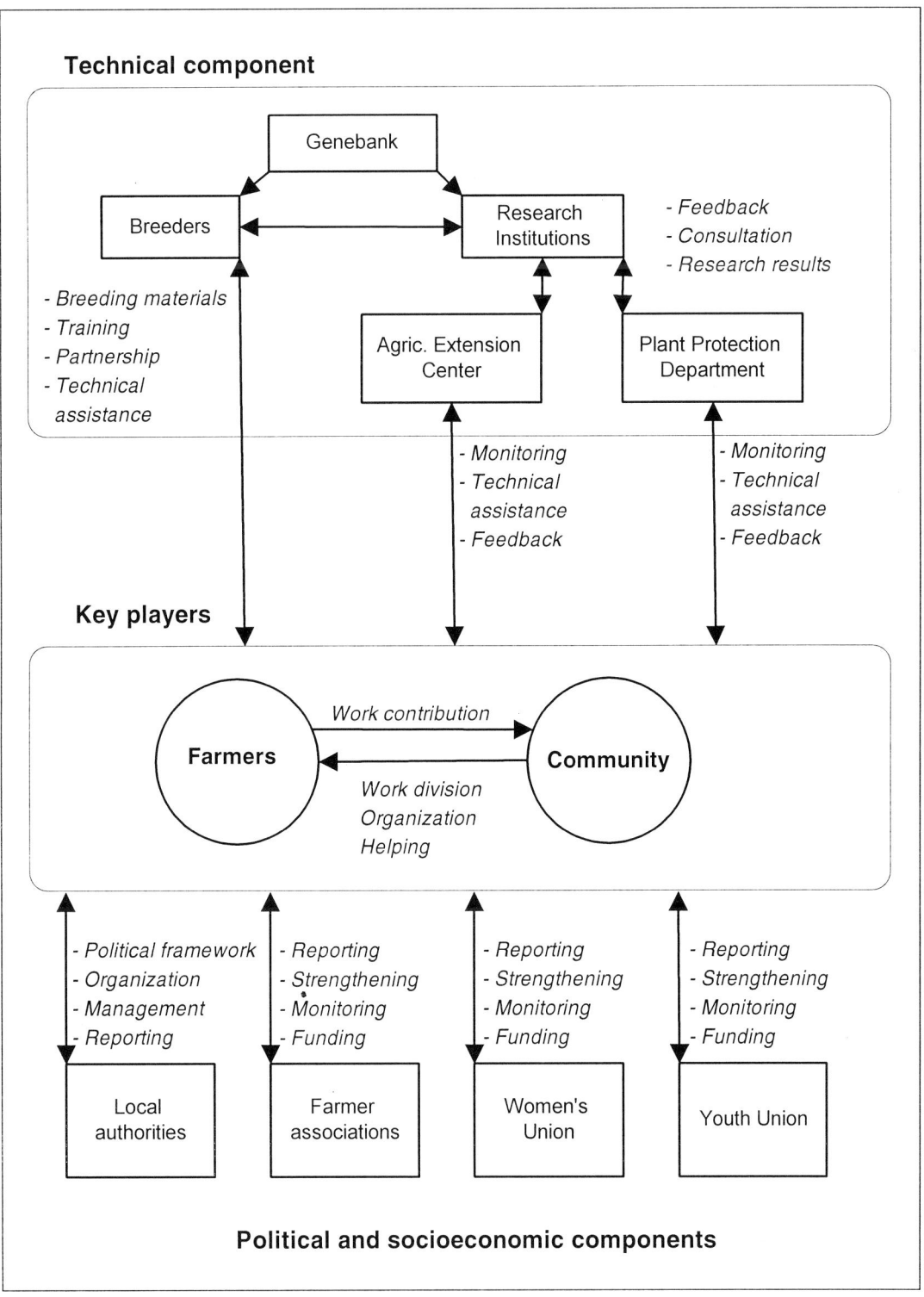

Fig. 7.2. The relationship between formal and informal sectors directly involved in Participatory Plant Breeding in the Mekong Delta.

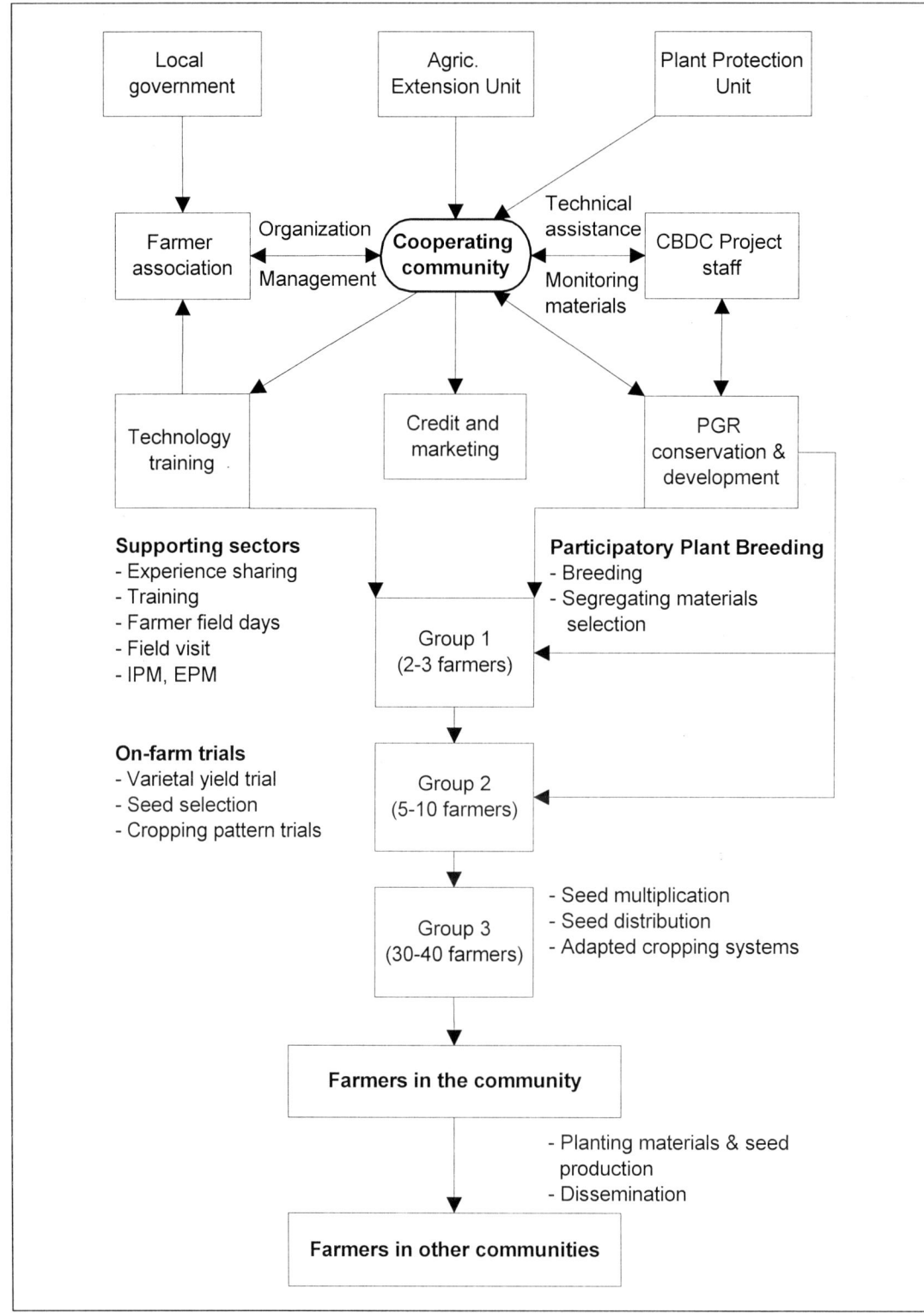

Fig. 7.3. Community-based networking diagram under CBDC project.

Selection of farmer's groups
One of the most important criterion for selecting farmer's groups is the willingness to cooperate based on mutual benefit and needs. The PPB/PVS programmes could not be successful and sustainable if the cooperation and participation were not voluntary.

Community meetings, field visits and personal communication with local extension workers and local organizations are the key tools for identifying cooperating groups of farmers.

Selection of farmers
Criteria for selecting cooperating farmers depend on a farmer's capacity in breeding and seed selection:
- Farmer "level 1" with willingness, good experience, skill and high potential in breeding and varietal selection
- Farmer "level 2" with willingness, experience in varietal selection
- Farmer "level 3" with willingness, experience in seed selection and multiplication.

As well, cooperating farmers should have a good socioeconomic identity:
- Relatively good economic condition, so he/she will have spare time for breeding activities
- Good technical knowledge and skill, so he/she can facilitate and manage all breeding activities at their communities
- Good social relationships, so he/she can help the project in organizing and grouping farmers.

Identification of farmers' needs and capacities
To identify farmers' needs and capacities, the following methods are used:
- Community meeting/workshop, and consultations with key informants and individual farmers are conducted to list their problems and needs on crop improvement; participatory prioritization of intervention is also made
- These events are used to identify a number of farmers with different experience, knowledge and skill in conducting crop-improvement activities through PPB/PVS on-farm
- Field visits are carried out to interview, observe, evaluate and discuss with listed farmers their interest and capacity in PPB/PVS implementation.

Farmer's capacity-building
Capacity-building for farmers is very important to improve farmer's practice and to assure the smooth operation in undertaking PPB/PVS by farmers. Some common but effective activities are provided:
- Training farmers on breeding techniques
- Organizing farmer field schools and farmer field days
- Undertaking some demonstration plots at each community
- Distributing technical documents such as manuals, leaflets, books, etc.
- Technical field assistance.

Review and planning meetings
Regular review and planning meetings at least every crop cycle provide good technical assistance and update information on the community situations and feedback from farmers to adjust the activities. These activities can better facilitate farmers' access to genebanks.

Implications and lessons learned
The results and achievements of these activities have had a strong impact on national seed policies, which are under review and reform in Vietnam. Experiences from the CBDC project have been used as proven events to convince policy-makers of the important role of farmers

and farming communities in the whole seed chain. Consequently, this approach has been adopted by many genebanks. More and more breeders and genebanks operate positively in linking directly with farmers, especially through PPB/PVS programmes. Farmers are now more aware of the existence and purpose of the national genebank. They can make seed requests or go directly to genebanks and research institutions to ask for assistance.

Linking national genebanks and farmers and farming communities can benefit both breeders and farmers. One should keep in mind that:
- Farmers conserve and develop PGR diversity for their needs and under the circumstances related to home consumption, market economy and adaptation to their specific environments, farming system and farm resources
- Linking farmers to genebanks makes genetic resources available for farming communities to improve their choice of genetic materials and better facilitate crop improvement in agriculture
- Support from local authorities and organizations in terms of organization, management, additional funds and facilitation is very important
- Cooperation with group/community on PPB/PVS will produce better results than with individual farmers
- Farmer Field School and Farmer Field days on PPB/PVS are good ways to motivate the farmers' participation at community level
- International, national and communal policies should realize the farmers' role and farmers' rights.

8. Toward establishing links between farmers and the ICRISAT genebank

Paula Bramel-Cox

Introduction

In the past, genebanks were established with various goals that impact the future conservation and utilization of their *ex situ* collections. Pistorius (1997) documented the history of plant genetic resources conservation nationally and internationally. The early goals and targets of collections were dominated by the needs of plant breeders. This had impact on the link of conservation with use, on the conservation of collections in specific regions in genebanks in Plant Breeding Institutes, and to the focus of many of these collections on crops of value to plant breeders. Pistorius (1997) concluded this also accounts for the focus on *ex situ* versus *in situ* conservation. These close ties between conservation and use by plant breeders also resulted in collecting missions that were organized by and for breeders. Thus, collecting strategies and information gathering focused on environmental sources of variation rather than cultural or consumption differences. Pistorius (1997) described this difference in approach as related to genecological ideas versus a more pure Mendelian genetics approach.

"For genecologists, landraces and wild relatives of cultivated plants – being the product of generations of conscious and unconscious selection by farmers and co-adaptation to their own habitats – represented a tremendous genetic reserve. Hence landraces were not valued so much for their specific genetic content, as for their ability to adapt to their environment while simultaneously developing resistance. As such, landraces were starting points…for breeding practices…as much as genetic donors"

Thus, genotypes or gene complexes found in plant populations needed to be the focus of future conservation programmes where emphasis would be on local adaptation and not single useful genes. Bennett (1970) suggested a method of collecting that used random sampling from farmers' fields which were selected based on biased sampling. She concluded that appropriate sampling of local variation required a consideration of climate and its local variants, soil type, topography, distribution of crops, history of cropping systems and varieties, knowledge of social structure, agricultural practices, and use of varieties by farmers. She felt that no collecting could be planned without detailed local knowledge of habitat variation within regions. This was not the prevailing view when most of the collections were made during the 1970s to 1990s at ICRISAT.

Rethinking in genebank: ICRISAT's experiences

The ICRISAT collections served mainly as working collections for plant breeders to evaluate and use to transfer specific traits. There was an early focus on direct use of collections as new varieties for farmers but that phase has passed for most of our collections. Thus, the primary user of our collection in the past and for the future is the research community. To enhance the direct use of the stored genetic resources by farmers, emphasis must be given to the description of local adaptation or the adapted gene complexes found in collected landraces and to better focus and document future collections using the approach suggested by Bennett (1970).

The traditional collecting trip at ICRISAT consisted of a team that included the expert on the crop from ICRISAT and a local expert. The collecting sites were selected based upon the knowledge of the crop specialist or priorities set by breeders of the crop. The focus was on collecting "valuable diversity" in landraces and/or wild species for use in the breeding programmes, by taxonomists for classification, or for conservation. A shift to a greater need for these collections as sources of local adaptation in breeding programmes, for local seed system restoration, and direct use by local communities for development will require a rethinking of the collecting procedures used and the types of information needed in the passport databases. The process – from identification of the collecting sites to the regeneration and evaluation of the collection – needs to be participatory with a broader group of users. The targeting of priority areas should include information on the environmental heterogeneity, distribution of the crop, history of the crop, cultural diversity, movements of people and possible threats of genetic erosion. There are many sources of this information and these need to be consulted in the planning.

Participation of local experts will be critical to this process and in the make-up of the collecting team. This team should have thorough knowledge of the possible sources of diversity among and within the landraces. Some understanding of the number of landraces grown in an area helps to predict the degree of diversity among landraces and ensures that all are adequately sampled. This information usually can be obtained from the local extension agent. Another source of this information may be in past crop surveys done by other government agencies, scientist or student thesis reports at the University, NGO's reports for food relief or community seed banks, and FAO or UNDP reports. A review of previous collecting trips and all germplasm that is currently held in other international or the national genebanks from that region would also be useful. The status of past collections by less traditional sources should be determined.

Rethinking the collecting strategy

The focus of collecting should be multi-purpose: crop diversity for the National Genebank, for developmental programmes, for specific breeding programmes, or for scientific study. All these need to be determined ahead of time and their specific requirements addressed. These could differ in the sampling strategy used and the information needed from the farmers. Sometimes these can be the same but it is best if they are viewed differently. Collecting for the genebank conservation would concentrate on maximum diversity with a minimum number of samples. The number of samples to be taken from a location or environment would represent the degree of diversity among all the factors. The samples would be harvested accounting for both among- and within-landrace diversity and in the fields. Collecting for the breeding programmes might concentrate on identifying sources of improved farmer varieties. This could be the same need as the development programmes but the information on their value from the farmers would be more important and targeted.

The planning phase should include specific requests for information from the NARS and the development of the team to represent the needs of the trip. The support in the country should include the research as well as the extension agency. Relevant NGOs should be involved in the planning, especially those with community seed bank experience, experience in rural appraisal methodology, experience in food aid, or any other development experience. This broad team for planning would then be narrowed for the actual collecting but their assistance may be solicited for local collecting activities. The planning should also include any reports developed from previous surveys or databases on crop-level diversity. These have to be studied to predict the degree of diversity in a locality and the possible urgency of the collecting. Priority areas will then be identified based upon all these sources of information and actual visits to targeted localities to set the final priorities.

Collecting can be conducted by a single team or it can be coordinated through local or regional staff. Each team should be made up of at least one women and local guides who speak all the local languages. The single team model was used in the past but requires more careful planning to time collecting to coincide with seed-harvesting time. This can present problems with the documentation because of conflicts of farmers' time and the range of maturities which may be grown. The best model might be to use a team to identify the specific collections to be made and then request local cooperation from the extension agents or NGO staff to make the final collecting of the seed. This model might require that special instructions be given these staff for seed drying or storage. If they obtain seed of acceptable quality from the farmer this can be accommodated. The farmers have a good understanding of seed conservation methods for their local conditions so seed samples obtained from their stocks will be of acceptable quality. Any special sampling strategies will need to be planned ahead of time and this information communicated to the farmers and their permission solicited for these special requirements. The supplies needed to sample individual plants or heads will need to be left with the collector. It is best to specify if you want individual heads or a sample of a bulk at the time you request the seed from the farmers. It might be best to be flexible to accommodate the procedure used by farmers to maintain their seed.

Documenting farmers' knowledge

The farmer survey used to better document traditional knowledge of the landraces should concentrate on five types of information:

1. Farmer name and description of environment. This can include the more traditional collecting site information and the farmer's own description of the field site. The farmer's names should be correctly documented.
2. Description of the variety characteristics by both the farmer and the collecting team. The use of simplified classifications to rapidly describe specific attributes will assist in both documentation and data collecting in the field. Statements that are more elaborate can be written in narrative on the forms as well. The general categories are morphological description, agronomic characteristics and specific stress reactions. Description should concentrate on both positive and negative aspects. This is some of the most difficult information to verify. From the farmer, an attempt should be made to only document this information clearly without limited questions on integrity. The collecting team should listen carefully to the farmer to learn from them, not to teach them or challenge the validity of their observations.
3. Description of the end use of the variety and its specific properties. This should include discussions of its cooking/ processing/ storing properties. The market value for specific products can be requested. The constraints to its use or storage should be asked about. The quality of its products can also be queried. The specific storage practices used for the crop and its seed should be asked about separately, as they can be very different.
4. Description of normal cultural practices used with this variety or landrace. This information is best taken in the field. It can include a comparison with other prevalent landraces grown by the farmers. A determination should be made of the roles each gender takes in the cultivation and an effort made to discuss the specifics with the proper household member. This can be difficult without a mixed-gender collecting team.
5. Description of the history of the variety with that farmer. This should include as thorough a discussion as possible of the varieties' selection history. If the question results in the identification of another farmer who had a long previous history then an effort may need to be made to collect from that farmer. The longer the history and the more effort made by the farmer to select the seed stock, the more important the collection. This question should also be explored in the field so the actual selection procedures used can be demonstrated by the farmer (either male or female). The actual gender of the selector should be determined for each phase of selection during harvest and storage.

The team should use a semi-formal questionnaire where the information can be verified by brainstorming after collecting is completed. A summary of all the information gathered can be written for the collecting trip report. All these documentation needs should be specified during the planning phase and incorporate the needs of all the participants.

Linking the genebank with the farming community: PPB

When the primary user of the collection is the National Programmes or the local communities, a broad partnership can be built with the NARS in conjunction with strengthening both the genetic resources and breeding programmes capability. Joint evaluations of the collected germplasm in the relevant local environments could be conducted. This grow-out could be used to regenerate and characterize the collection in the country of origin for the genebank. A field day could be planned to solicit farmers' input into the identification of useful locally adapted farmers' varieties that could be the basis of the NARS breeding programme. This collecting, documentation and evaluation of locally adapted farmers' varieties could result in the design of a participatory breeding programme that will be better able to meet the needs of these traditional farmers. This would allow the primary benefits from this collection to be shared among the country of origin and the local communities. The global importance of the collection would be secondary but its value would be enhanced with the additional information.

This broader, participatory approach will result in an *ex situ* collection that can fill the needs of the farmers for secure conservation of their germplasm. It also enhances the understanding of the international and national genebank and plant breeding programmes of the needs for farmers of locally adapted but better-yielding varieties. The interaction of farmers with the present ICRISAT collection requires a different approach to the evaluation and distribution of germplasm. The existing collection does not have the documentation that was described for future collections. In fact, very little of the ICRISAT collections are documented with any of the information given as essential in the procedures described earlier. A careful re-evaluation of the previous collecting reports may recover some of the information but this will not assist in documenting collections that have been acquired through donations. One might conclude that the present collections held by ICRISAT *ex situ* are of very limited use for the farmers who contributed to its unique properties. Farmers will be mainly interested in evaluating particular types, although there could be some interest in new variation and specific traits by farmer participatory plant breeding programmes. These needs should be addressed by encouraging greater interaction of NGOs, communities or farmer groups with the collections. At ICRISAT, this could mean targeting distributions of specific germplasm for evaluation by these groups or by enhancing the information flow about the collections to these groups. This local use of genetic resources is more difficult to target but will depend upon good communications and broad partnerships within a region or country.

For meeting the needs of farmers for specific local varieties or types, the existing information on the collections is inadequate in most cases. The farmers are interested in the recovery of varieties or crops that have suffered from genetic erosion. This loss could be due to natural disasters, war, famine, changes in cropping systems, or any other social or marketing factors that have resulted in rapid shifts to new varieties or crops. Short-term changes in crop choice or the epidemic of a new disease or pest can have a significant effect on the availability of local landraces in the long term. This vulnerability of the local seed-supply system is particularly a problem for the relatively minor crops that are the focus of our collections. The need to restore or repatriate germplasm to the specific region, community or farmers requires the type of information that will be collected in the future. Thus, meeting this need is difficult with the existing collection. Also the collections have been through a number of regenerations outside the local environment and the diversity has been redistributed from within to among accessions. This could result in genetic shifts in the accession that may influence its adaptation to the local condition or its stability, owing to the loss of population buffering. While we have met requests for specific landraces from specific collecting sites, these have not been well received by farmers with prior knowledge of the landrace or a familiarity with the properties of the original landrace.

Two activities would enhance the opportunity for ICRISAT to contribute to these restoration needs. One is to work with the farmer or community to introduce the most closely related material, both from the region and from closely related environments. A second activity is to conduct a careful evaluation of potentially acceptable germplasm with the farmers. The insight gained from this exercise would be added to the existing database at ICRISAT to enhance its use for the future and would add to the farmers' understanding of the range of variation available in specific genetic material. The most acceptable germplasm would be increased, distributed and tested at the local level for use by the target farmers. This evaluation of germplasm *per se* was carried out by the Genetic Resources Program in the past but exclusively with agricultural researchers.

One other important use of our collections by farmers could be as sources of improved varieties or new sources of traits. One case which has been requested is the need by a certain region or community to have earlier maturity varieties in the cropping system. All the locally adapted landraces are late maturing and the high-yielding improved varieties lack specific quality traits. Locally adapted varieties with earlier maturity could exist within the collectiond from neighbouring regions or closely related regions. Again, the identification of local landraces to introduce to the region for direct exploitation by farmers would be desirable. A careful evaluation of our characterized collections and our limited information on adaptation could be used to target germplasm for these trials. The farmers in the region would be responsible for further use of the germplasm but the information gained on our collection would be useful to meet other requests. Hodgkin and Anishetty (1999) described some of the needs for restoration of local seed systems. That conference, held in 1999 by FAO, concluded that greater access to information and genetic material from local regions is a key to restoration of local seed systems. This information needs to be strengthened and made more accessible to meet this increasingly important need.

Back to the farmer: access to germplasm
Enhancing access to the ICRISAT genebank by farmers has required a change in the type of information we gather during collecting trips, in our conservation strategy to minimize regeneration and maintain the genetic integrity of the original sample, and in the use of information to target germplasm requests. There are opportunities to use the existing collections to meet these needs but only in closer cooperation with the region or community. The enhanced use of our collections by farmer breeders or participatory plant breeding programmes will also require a different strategy from that used by the research community. The main reason for this is the greater interest by farmers in local adapted complexes as the parents for improvement. ICRISAT, like other CGIAR Centres, has seen the primary user of its collections as researchers, especially plant breeders. While they will continue to be our primary users in the future for the majority of our crops, the need by farmers for genetic resources of these minor crops will have to be addressed. The marginal nature of the crops we conserve and the environments in which they are grown means the need for the *ex situ* collections held by us will increase as the uncertainty of the environment increases. The need to focus effort on the direct improvement of production in these regions or to enhance diversity with increased productivity of minor local species will be addressed efficiently with a greater use of the collections *per se* by farmers. This focus has required ICRISAT to look at documentation, evaluation and even conservation differently for the future.

References
Bennett, E. 1970. Tactics of plant exploration. Pp. 157-129 *in* Genetic Resources in Plants: Their Exploration and Conservation (O.H. Frankel and E. Bennett, eds.). IBP Handbook No. 11. Blackwell Scientific Publishers, Oxford, UK.

Hodgkin, T. and M. Anishetty. 1999. Plant genetic resources and seed relief. Pp. 139-146 *in* Restoring Farmers' Seed Systems in Disaster Situations. Proceedings of the International Workshop on Developing Institutional Agreements and Capacity to Assist Farmers in Disaster Situations to Restore Agricultural Systems and Seed Security Activities, Rome, Italy, 3-5 Nov. 1998. FAO, Rome.

Pistorius, R. 1997. Scientists, Plants and Politics – A History of the Plant Genetic Resources Movement. International Plant Genetic Resources Institute, Rome, Italy

9. Re-introducing crop genetic diversity in post-war Somalia

Esbern Friis-Hansen and Dan Kiambi, with Luigi Guarino and James Chweya†

Emergency seed supply

In 1995, 26 countries experienced complex humanitarian emergencies such as natural events (drought, flood, cyclone, earthquakes, etc.), man-made crises (civil war, failing economic, social and political institutions) or a combination of both. An estimated 37 million people, more than half of whom were Africans, were deemed to be in need of aid to avoid starvation (Scowcroft and Scowcroft 1997).

The FAO Global Plan of Action on Plant Genetic Resources for Food and Agriculture states: "In the modern world and especially in developing countries, people are threatened with and vulnerable to natural disasters, civil strife and war. Such calamities pose huge challenges to the resilience of agricultural systems. Often, adapted crop varieties are lost and cannot be recuperated locally."

Historically, emergency aid organizations have provided large quantities of seed to farmers in emergency situations. However, such seed is most often imported and not adapted to local growing conditions (pests, diseases, cultivation systems and food preferences). Food aid combined with emergency seed assistance supplying poorly adapted seed may result in loss of local crop diversity and have a negative impact on household food production and security in subsequent years (Friis-Hansen and Rohrbach 1993; Richards and Ruivenkamp 1997; Sperling 1997; Friis-Hansen and Kiambi 1998).

Crop diversity, both in terms of number of crops and number of varieties of each crop, plays a crucial role for resource-poor farmers for whom household food security is a major production goal. Crop diversity (1) allows farmers' cropping systems to adapt to local ecological microniches of their particular fields, (2) satisfies household food preferences, and (3) provides tolerance to pathogens. If farmers lose their crop diversity as a result of the combination of catastrophe and poorly adapted emergency aid, their subsequent ability to grow sufficient food may be seriously negatively affected. While addressing an immediate crisis, emergency aid can exacerbate conditions of hunger and undermine food security in the long run.

A number of UN agencies and international agricultural research institutes and NGOs have recently gained experience with restoring locally adapted weed and planting materials following emergencies in Southern Africa, Rwanda, Sierra Leone, Somalia and Cambodia. These experiences show a wide variety in the nature of emergencies and their impact on crop diversity. Further research and analysis of different types of emergencies, the variation in resilience of local seed maintenance systems and how emergency assistance can best be designed in order to address the immediate needs of a community who have undermined its longer term self-sufficiency, is required.

Identification of development requirements and participatory methodology in Somalia

An unnoticed impact of the sociopolitical strife in Somalia was the destruction of the research and conservation facilities in Baidoa/Afgoi in 1992, during which the national germplasm collection was destroyed. The state of civil war continued, leading to the total collapse of state institutions and enormous destruction of modern physical structures. By the mid-1990s, it was clear that the agricultural production system was adversely affected and this had led to loss and/or reduction of genetic diversity *in situ*.

As a response to this grave situation, IPGRI embarked on multiplication and characterization of 284 accessions of sorghum and maize germplasm which were conserved as a duplicate collection at the National Genebank of Kenya, with the intention of repatriating it to Somalia for use in the agricultural production systems. In order to achieve this goal, a project on "Re-introduction of germplasm for the diversification of Somali agricultural production system" was formulated by IPGRI, in partnership with CINS (an Italian-based NGO), which received funding by the European Community. The project objectives were to (1) sustain and increase the genetic base of Somali agricultural production systems through re-introduction of germplasm, (2) create awareness of plant genetic resources conservation among farmers and scientists in Somalia, and (3) strengthen seed production and distribution capacities through training and development of linkages.

Such an exercise had never been undertaken before and the methodology was developed as the project proceeded. The participatory approach aimed at creating a partnership between the project and farmers, and their legitimate social institutions at the village level.

The project seed centre was attacked by a local militia in November 1997. A Somali guard was killed, several wounded and the Italian NGO staff was kidnapped for three days. The project seed centre was totally destroyed, including the *ex situ* conserved varieties and all records. The project was subsequently discontinued and this chapter will therefore be limited to a discussion of the participatory methodologies used, as it is not possible to discuss the impact of the project on farming communities.

Dialogue between farming communities and scientists

Initial meetings were held with village elders during which the project objectives and planned activities were explained. The farmers responded positively and offered advice and information on the prevailing socioeconomic and crop production constraints. Then followed a series of group meetings in three villages, during which farmers discussed the project among themselves. Farmers were also requested to raise additional issues to be addressed. These dialogue meetings included the village chairmen, respect-ed village elders and men and women who were skilled in managing local crop varieties.

The dialogue meetings started with a session during which the group of village elders discussed the major elements of their farming systems that included land tenure, labour market and effects of the civil war on their livelihood. After gaining initial knowledge about the general social and agricultural context, discussions were guided toward the indigenous plant genetic resource management practices, including selection, storage, exchange and use of crop varieties (Fig. 9.1). A smaller group of farmers, including women, thereafter answered a semi-structured questionnaire designed to obtain farmers' views on agronomic characteristics of each of their local crop varieties. The results of these participatory characteristics of crop species are shown in Table 9.1.

Assessment of farmers' post-war socioeconomic situation and germplasm needs

In order to gain an understanding about the social differentiation within each village, a participatory wealth ranking exercise was carried out. A small group of knowledgeable farmers was asked to agree on a number of qualitative criteria for differentiating farmers within the village into three wealth categories. In all the villages, these key informants were able to agree on a common set of criteria with only minor disagreements. The criteria varied from village to village and the wealth categories are therefore relative to the local context. Table 9.2 shows an example of locally formulated wealth ranking criteria.

Two-thirds of the families in Casirmaal village are permanently food insecure. Following two years without any food relief, some people in this group were so weak that they were unable to do hard work in the field. As farmers in this group work for others to survive, they are not likely to be able to break out of their poverty trap without targeted outside assistance. Only the wealthiest (10% of the farmers) are food and seed secure. These few families do not

Enhancing farmers' access to PGR

Fig. 9.1. Village elders discuss their seed requirements with scientists from IPGRI.

Table 9.1. Farmers' assessment of available maize varieties in Kongo and Shameento villages †

	Name of maize variety			
Characteristic	Local I	Local II	Samtux	ICRC
Special character	up to 23 rows of seed per cob	up to 18 rows of seed per cob	national research programme	emergency seed distribution
Grain yield	high	high	medium	low
Grain yield potential	average	average	high	?
Grain size	large flat	small round	small	large flat
Head size	medium	large	medium	medium
Drought tolerance	average	average	poor	good
Time to maturity	90-100 days	90-100 days	110-120 days	70-80 days
Water requirement	3 periods of irrigation	3 periods of irrigation	4 periods of irrigation	?
Non-grain yield	average	average	good	low
Stem quality	good fodder	good fodder	poor fodder	?
Disease resistance	average	average	?	poor
Insect resistance	average	average	poor	poor
Threshing ease	difficult	easy	easy	easy
Grain taste	good	good	good	good
Grain storage	good	good	poor	poor
Demand for fertilizer	average	average	high	high

† Information is based on farmers' knowledge and perception. The column headings are farmers' own names for existing maize varieties.

have sufficient surplus during periods of crop failure to share with other families. Their grain pits are buried at secret locations and only opened in privacy during the night.

While the participatory dialogue and wealth ranking had generated much knowledge, a clear need for a quantitative assessment of the situation emerged. A socioeconomic and farming systems baseline household questionnaire survey was conducted as a response to this need, covering 266 households in nine villages. On the basis of the initial dialogues with farmers, along with project documents, a literature review and NGO consultations, a draft questionnaire was developed, tested and reviewed with farmers in Somalia.

Because of the security situation in Somalia, the baseline survey was carried out by a Somali agronomist employed by the project. The following procedures were followed to select which farmers should participate in the survey. First, an inventory of all adults living in each of the villages was conducted, either on the basis of a pre-civil war list, or on the memory of the village chairman and elders. Second, 30 farmers were chosen at random from the inventory list to participate in the socioeconomic and farming system baseline survey. Wealth-ranking criteria were applied on the selected farmers, grouping them into three wealth categories.

The questionnaires were entered into two SPSS databases (on a household basis and on a field basis). The data were converted from local units (*taab* and *jibal*) to hectares. A code book was generated and statistical analyses on a household and a field basis were carried out. District and village profiles were developed based on information on the village-based checklists. Data on characterization of all local sorghum and maize varieties based on farmers' perceptions were compiled. A hard copy of the entire socioeconomic and farming systems baseline database was compiled.

Transfer, testing and selection of germplasm

The National Genebank of Kenya played a crucial role in the initial phase of the project through multiplication, characterization and testing the viability of the genetic materials, and assistance in packaging and procurement of the phytosanitary certificates. The Genebank is also conserving a duplicate set of the accessions for safety.

The germplasm previously stored at the Kenya National Genebank was air-freighted by a chartered aircraft to the CINS seed centre located in Decalay village in Balaad district, Middle Shaabel, Somalia. The material weight about 0.5 tons and the total number of accessions was 165 (151 sorghum and 14 maize accessions). The genetic materials were subdivided into four parts. Some were stored in freezers run by a generator with another generator as back-up; some were distributed for on-farm trials; some were distributed for trials at the Decalay village Seed Centre; while the last part was cultivated under a high-input regime for the purpose of multiplication.

On-farm trials

The set for on-farm trials was subdivided among the nine villages located in the three districts in which the project was being implemented. The villages specialize in growing either maize or sorghum. One village grows both maize and sorghum. Two sorghum-growing villages and one maize-growing village were selected in each district. Ten farmers in each village were selected for participation in the project using the rapid rural appraisal approach. The sorghum villages received 20 accessions each (17 different and 3 accessions common to all). The material chosen represented all seed colour types. The three maize-growing villages received 10 accessions of each. One village received both maize and sorghum accessions. The sorghum farmers received two accessions each while the maize farmers received one accession each (Fig. 9.2).

Ten farmers were selected for on-farm trials in each of the three villages in each of the three districts in which the socioeconomic and Farming Systems baseline survey was carried out. Two to three varieties were given to each of the 90 farmers to cultivate during the 1997 *gu*

Table 9.2. Results of participatory wealth ranking in a Casrimaal village [†]

Low (65%)	No sorghum grain stored in grain pit Less than 1 ha cultivated with sorghum No long-term storage of seed No livestock Members of household work as casual labourers for others
Middle (25%)	Sorghum grain stored in grain pit is insufficient to meet household requirements in case the seasonal rains fail 2-4 ha cultivated with sorghum No long-term storage of seed 1-2 dairy cows 1-3 goats Household members do not work as casual labourers unless they run out of food
High (10%)	Sorghum grain stored in grain pit is sufficient to meet household requirements in case the seasonal rains fail 5-30 ha cultivated with sorghum Long-term storage of sorghum seed 4-5 dairy cows 10-15 other cattle herded by hired herdsman 3-5 goats Employ others as casual labourers

[†] Information is the result of a participatory wealth ranking based on farmers' own holistic views of social differences within the village.

Fig. 9.2. Somali farmer tests germination of stored seed by planting the seed in moist soil surrounding a clay pot. The seeds are only used if more than 8 out of 10 germinate.

season. The farmers were asked to grow the varieties in a corner of their own field and use their traditional cultural practices. Three household questionnaires were used to record information from the participating farmers during different stages of the 1997 *gu* growing season.

On-station trials

The on-station trials had dual purposes: (1) to characterize all sorghum, maize and sim-sim varieties to complement the characterization carried out by the Kenyan Genebank, and (2) to carry out participatory on-station farmer selection of desired varieties. Each variety of germplasm was grown in a 2 x 2-m area. Adjacent to the on-station trials, larger plots were cultivated for the purpose of multiplying material of which there were sufficient quantities of seed. Agronomic characterization of all varieties was carried out by CINS staff and standard protocols for agronomic characterization were applied. The protocols for the participatory characterization were developed in consultation with participating farmers. It was planned that 15 farmers from each of the three districts would be invited to the on-station trial twice, during flowering and harvest, in the 1997 *gu* growing season. An elaborate system had been developed for farmers to select their preferred varieties, based on both individual and group assessments, and on answers from both men and women. Unfortunately, because of an upsurge of intermilitia fighting, farmer's participation in the on-station variety selection was cancelled for security reasons. The experimental plots were severely flooded at the end of the *gu* growing season which resulted in serious crop loss. Because of the resulting low yields, the purpose of seed multiplication was only satisfied to a limited extent.

Multiplication and distribution of selected germplasm

The baseline survey revealed that all farmers are interested in new varieties and most are willing to try out such varieties on a limited scale. The on-farm trials, moreover, showed that there is considerable interest among farmers in adopting the varieties to be produced by IPGRI/CINS. As a response to this, IPGRI developed an appropriate methodology for multiplying and marketing seeds under the current conditions.

Currently, there are no functional state institutions in Somalia, and farmers rely on their own resources, as well as on the private sector and donors for agricultural services. This situation is not likely to change in the immediate future. If provision of a service such as supply of seeds is to be financially and technically sustainable, i.e. to continue after project completion, it must be based on the private sector and local community institutions. The approach of the project, therefore, was to improve and build on existing knowledge about management of plant genetic resources and work with local institutions and individuals. The CINS Seed Centre was planned to provide foundation seed and technical support to the 30 village-based seed farms and to sell seed to farmers through 15 marketing women operating in each of the three district towns.

The CINS Seed Centre was perceived to store all varieties under secure conditions. The village seed farms were planned to be located throughout the three districts and each farm would ideally cover seed requirements of five villages each. A village seed bank was planned to be established in connection with each of the village seed farms with the aim to store small quantities of all varieties cultivated in the area and store adequate quantities of the most popular varieties to serve as a buffer in case of crop failure and widespread loss of seed. The village seed farms would, in theory, become self-sufficient with seed after one season. However, it is likely that if the project resumes when peace has returned to Somalia, there will be a need for periodic supply of high-quality seed of specific varieties.

Conclusion

The attack on the Decalay Village Seed Centre by a rival militia group had a disastrous effect for the outcome of the IPGRI/CINS efforts to re-introduce germplasm to farmers in post-war

Somalia. The Decaley Village Seed Centre with *ex situ* seed and *in situ* conservation was destroyed; the planned and imminent participatory variety selection process was disrupted; the seed multiplication plots were never harvested and planned seed distribution activities were abandoned. This is the inherent risk of seeking to work with development in a volatile post-war but pre-peace situation. As part of the participatory approach, the full diversity of germplasm was tested in 90 on-farm trials at the time of the militia attack. While the results from these trials were never analyzed, the planting material survived in farmers' fields and preferred varieties may spread through community seed exchange.

There are nevertheless important lessons to be learned from this project. First, the participatory approach with a strong social science component used by IPGRI in this project is novel and unique for emergency seed programmes and could serve as inspiration for future programmes. Second, IPGRI's approach to providing access, transferring, multiplying and testing a wide range of *ex situ* stored indigenous varieties, instead of modern varieties from other countries (which is the case of many other emergency seed programmes), was very well received by Somali farmers and should be an element in future emergency seed programmes when possible.

References

Friis-Hansen, E. and D. Kiambi. 1998. Re-Introduction of Germplasm for Diversification of the Genetic Base of the Somali Agricultural Production System, IPGRI Technical Project Termination Report. IPGRI, Rome.

Friis-Hansen, E. and Rohrbach. 1993. SADC/ICRISAT 1992 Emergency Production of Sorghum and Pearl Millet Seed. Impact Assessment. ICRISAT Southern and Eastern Africa Region Working Paper 93/01. ICRISAT, Hyderabad.

Richards, P. and G. Ruivenkamp. 1997. Seeds and survival: Crop genetic resources in war and reconstruction in Africa. IPGRI, Rome.

Scowcroft, W.R. and C.E.P. Scowcroft. 1997. Seeds security: Disaster response and strategic planning. Workshop paper. FAO, Rome.

Sperling, L. 1997. War and Crop Diversity. AGREN Network Paper No. 75. ODI, London.

Section III

Local plant genetic resources management and participatory crop improvement

10. Overview

Bhuwon Sthapit and Esbern Friis-Hansen

Local plant genetic resources management

This section presents some innovative ways of managing and also improving local agricultural biodiversity at a local level. Farming communities maintain agrobiodiversity because it is essential to their survival. Agricultural biodiversity which naturally exists on-farm results from an ongoing evolutionary process managed by farmers and shaped by the heterogeneity of environmental and social conditions in which they live. Farmers select and breed new cultivars for their use; thus, for farmers the concept of conservation *per se* may not be explicit. Farmers will continue to grow and preserve seeds of local biodiversity as long as they derive benefits in growing them. Agrobiodiversity will not be saved unless it is used. If the continued use of local cultivars by farmers is to form part of a conservation strategy, it is important to understand why farmers grow landraces, when they grow them and how they maintain and use them.

In situ (on-farm) conservation has, therefore, been proposed as a strategy to conserve the processes of evolution and adaptation of crops to their environments. It is also a continuation of the social processes of local plant genetic resources management among farmers and therefore is a dynamic process. Its purpose is to ensure the continuous adaptation of local genetic material to environmental microniches and the changing socioeconomic conditions of agricultural production. *In situ* conservation of plant genetic resources is not simply a question of conservation of varieties at a given location, but is, more importantly, aimed at ensuring the process of continuation of local plant genetic resources management practices and the maintenance and proliferation of the local knowledge that guides these processes. *In situ* conservation has the potential to (1) conserve biodiversity at all levels – the ecosystem, the species and the genetic diversity within species, (2) improve, maintain or increase control over and access to the genetic resources by farmers, (3) improve the livelihood of resource-poor farmers, and (4) integrate farmers or informal sectors into the national PGR system of conservation and use (Jarvis *et al.* 1998).

In situ conservation has also been suggested as a strategy, complementary to *ex situ* conservation, to conserve intra-species crop diversity among crops cultivated by farmers (Altieri and Merrick 1987; Brush 1991; Eyzaguirre and Iwanaga 1996; FAO 1996; IPGRI 1996). In recent years, a number of NGOs and international and national PGR programmes have responded to calls for increased local participation and *in situ* conservation by initiating on-farm conservation projects. On-farm conservation of crop diversity is only possible through the active participation of farmers and may not, therefore, be implemented by plant genetic resources scientists and technicians alone. Participatory approaches are crucial for the successful implementation of *in situ* conservation programmes because the actual crop activities are carried out by farmers and any change to local plant genetic resources management practices requires farmers' full support and active participation. Participatory approaches are also required in order to permit outside project staff to enter into a dialogue with farmers and become familiar with their indigenous knowledge and the basis for making decisions with regard to plant genetic resources conservation and use.

The need for outside assistance (from NGOs or national plant genetic resources programmes) to ensure the continued practice of *in situ* conservation by farmers within a particular area varies greatly depending on the particular situation and the objectives and nature of the development organizations. *In situ* conservation programmes are likely to target areas (1) which are known to contain a high level of genetic diversity, and (2) where such diversity has come under threat of genetic erosion.

In situ conservation programmes are likely to take different forms depending on the socioeconomic and ecological contexts. However, *in situ* conservation projects usually include activities that rely on participatory approaches. These include: (1) awareness-raising about the importance of plant genetic resources, (2) support of community management of plant genetic resources such as diversity fairs and community biodiversity register, (3) participatory plant breeding, (4) establishment of community-based small-scale *ex situ* genebanks, and (5) research.

In 1996, IPGRI, in collaboration with a number of national programmes, initiated a research programme implementing different approaches to *in situ* conservation. A key research issue in this programme is to study the inter-relationship between the availability of genetic variation within local crop varieties and local farmers' decisions about planting, managing, harvesting and processing their crops. The objective is to strengthen the scientific basis of *in situ* conservation of agricultural biodiversity on-farm (IPGRI 1996; Jarvis and Hodgkin 1998).

Participatory Plant Breeding

Many factors influence the availability of varieties to a farmer or a group of farmers at a given location, and the level of intra-species crop diversity varies greatly among geographical regions as well as locally. Farmers in poor communities do not necessarily have access to a wide spectrum of intra-species crop diversity developed over centuries. Most communities are situated outside centres of crop diversity and major crops in a given area may not have a long tradition of cultivation. There are two major sources for access to new crop varieties: (1) seed retention or community seed exchange, and (2) formal organized seed production and supply. Both of these sources of new seed are functioning poorly in many parts of the world.

During the 1980s, a number of NGOs and scientists working with agricultural development began to notice a significant divergence between conventional plant breeders' criteria for determining the acceptability of new plant types and the criteria used by poor smallholder farmers cultivating under rain-fed conditions, particularly those living in what have been called complex marginal areas. It became increasingly clear that this group of poor farmers had valid reasons for rejecting modern varieties. As a consequence of this realization, some plant breeders began to invite farmers to evaluate pre-released varieties on-station (Maurya *et al.* 1988; Salazar 1992; Sperling *et al.* 1993; Sthapit *et al.* 1994). Participatory approaches to plant breeding have since expanded greatly and are today used (although often still on a limited scale) at all major international agricultural research centres, in many national crop breeding programmes, and among numerous NGOs and CBOs (PRGA 1999). So far PRGA has documented a total of 48 case studies on participatory approaches to crop improvement (Weltzein/Smith *et al.* 1999) and this number is reported to be growing very quickly. Sthapit *et al.* (1996), Joshi and Witcombe (1996) and Witcombe *et al.* (1996) published three case studies in which full PVS and PPB processes were well documented. The product of PPB has been officially released and is spreading from farmer to farmer (Joshi *et al.* 1997; Witcombe 1997). Chapter 11 of this report describes PVS and PPB methods that successfully generated farmers' variety Machhapuchre-3, with the collaborative participation of local farmers in the high hills of Nepal. Farmers were involved in setting breeding goals as well as selecting segregating materials in their own plots and multiplying and exchanging seed among themselves. Breeder assisted in multilocational testing and the variety release process. Many PPB products that are preferred initially by a few farmers have now spread through informal seed networks.

As participatory plant breeding has become widely accepted and practised by formal plant breeding institutes, breeders have begun to involve farmers even in early stages of the research process (Iglesias and Hernandez 1996; Ceccarelli *et al.* 1996; Sthapit *et al.* 1996; Witcombe *et al.* 1996). Many participatory plant breeding (PPB) programmes are still experimental and are exploring new forms of participation and division of responsibility between scientists and farmers, including involving farmers in evaluating advanced breeding lines both on-station and on-farm, selecting from segregating plant populations, and participating in decisions as to which varieties to multiply and disseminate (PRGA 1999). The case from Vietnam (Chapter 12) also demonstrates that participatory approaches to crop improvement have been useful in high yield potential areas of Mekong Delta where IRRI varieties have made good impact. PVS in high yield potential systems in India and Nepal also provides convincing evidence that participatory approaches in variety selection in high yield production systems are equally effective (Witcombe 1999). The case demonstrates that it is not valid to argue that, because wide adaptation can be exploited in High Potential Production Systems (HPPSs), farmer participatory research is only appropriate for marginal environments.

The bulk of PPB is carried out by plant breeders who reach out from their research stations and involve farmers in the plant breeding process. Farmers may be involved in on-station trials or in trials in farmers' fields. While in most PPB programmes the research agenda is almost always set by the researcher and the experiments designed by scientists, a growing number of PPB programmes take a different approach and provide scientific and logistic support to farmer's own plant breeding. Such programmes are called scientific-aided indigenous plant breeding (Richards 1996) or farmer-led PPB (PRGA 1999) and have the following two major characteristics: (1) they seek to improve local indigenous plant breeding techniques and thus sensitize farmers to the value of plant genetic resources, and (2) they provide farmers with access to enhanced intra-species diversity (i.e. early generation plant populations derived from crosses of local varieties).

Witcombe (1997) suggested two types of PPB: consultative and collaborative. In consultative PPB, farmers are consulted at every stage to set breeding goals and choose parents that are appropriate to local conditions. In collaborative PPB, farmers grow the early, variable generations and select the best plants among them on their own fields. The choice of consultative and collaborative PPB will depend upon the crop and the availability of resources. Sthapit et al. (1996) demonstrated that collaborative PPB was practical in marginal areas where farmers were desperate for new options whereas the case may be different in HPPSs in Chitwan, Nepal where consultative PPB may be enough (Witcombe, pers. comm.). Therefore, PPB covers the full range of genetic improvement activities; setting breeding goals, creating genetic variability, selecting within variable populations, evaluating experimental varieties and producing new seed and disseminating from informal and sometimes formal seed networks.

There is no one method of PPB. The key element in PPB is a systematic blending of farmers' knowledge, skills and preferences with plant breeding principles and goals. The flexibility of the definition is important because degree of participation depends upon capacity and interest of collaborating farmers and researchers (including breeders), resources and institutional context. In this case, PPB can be defined in its broadest sense, ranging from plant breeder controlled decentralized breeding to various degrees of farmer involvement in the breeding or improvement process. Local crop development (informal breeding) by farmers and formal plant breeding by professionals in seed companies and research institutes are, despite different systems, based on fundamentally similar genetic and plant breeding principles. Degree of sophistication in techniques may vary, however. Chapter 13 of this report documents the value of a simple breeding method (mass selection) used by farmers of Nepal and Mexico in local crop development of maize and rice crops. This form of breeding has a significant role in those areas of developing countries where retention from farmers' own harvests and farmer-to-farmer seed exchanges are still the most common sources of quality seed.

PPB approaches thus draw upon the comparative advantages of both the formal and informal systems. In recent years, PPB also has been considered as a potential strategy for enhancing biodiversity and production. In this case it is important to distinguish between PVS and PPB as two distinct processes. PPB and PVS are both part of a breeding process in which farmers select cultivars under target environments. The only differences are: PVS is the selection of finished products by farmers in their own fields whereas PPB is the selection of segregating materials at an early stage in their own fields and/or bulk of heterogenous materials with an objective of generating a new composite variety (Sthapit et al. 1996). Understanding this distinction is technically important because the effects of PVS are considered unpredictable as it results in either an increase or a reduction in biodiversity at a local level. In contrast to the unpredictability of PVS, PPB is considered to have a generally favourable impact on biodiversity and could be considered a dynamic form of *in situ* conservation (Witcombe et al. 1996; Witcombe 1999).

By definition, PPB has significant farmer participation but it also involves decentralization of the breeding process from research station to farmers' fields. This may also speed up the process of adoption of new variety and increase site-to-site hidden genetic diversity. In this way, PPB is used to deploy diversity in farmers' fields for sustainable ecosystems and also to strengthen the process of on-farm conservation. Planting, managing, selecting, storing and exchanging materials are integral parts of such a process of on-farm crop conservation, and farmers' decisions play an important role in this process. The case study presented by Rana *et al.* (Chapter 14) shows participatory approaches to understanding farmers' decision-making processes at plot as well as community levels.

In 1997, the IPGRI-supported *in situ* conservation project also employed PPB as a method for adding benefits to on-farm genetic diversity. In this sense, PPB may play a role in 'adding value' to local landraces, increasing their chance of being maintained by the farmers. In order to maintain and use landraces, crop genetic resources must (1) be competitive with other options a farmer might have, and (2) contribute to the security and possible increase in a farmer's income (Jarvis et al. 1998). There are two main ways of adding value or benefit to crop resources: first, the material itself may be improved by adding value in terms of use, economic return, etc., and second, the demand of the material or some derived products may be increased by providing market or non-market incentives, policy relief and awareness of ecological and environmental benefits. The first option is to identify reasons for genetic erosion of a particular landrace and then seek to eliminate undesired traits through participatory plant breeding (see Chapter 11 by Subedi *et al.*). The second option includes adding benefits to local diversity through non-breeding means such as better processing, packaging, promoting and marketing (see Chapter 22 by Rijal *et al.*). A PPB approach which uses landraces as the source of genetic material for crop improvement symbolizes a balance between the two goals of maintaining genetic diversity *in situ* and improving varieties according to the needs of farmers. If landrace diversity as genetic resources is going to be conserved on-farm, it must happen as a spin-off of farmers' productive activities and for this PPB is seen as a potential tool.

While it may be too early to evaluate the success of participatory plant breeding, evidence from recent case studies suggests that varieties produced through participatory plant breeding and participatory variety screening programmes are well adapted to specific environmental conditions and the socioeconomic circumstances of the relevant population. It appears that varieties developed on the basis of the participatory model both increase participating farmers' yields and satisfy a range of non-grain yield production goals. It also appears that farmers who, as a result of their involvement in PPB have access to more diversity, tend to select and use more varieties.

References

Altieri, M.A. and L.C. Merrick. 1987. *In situ* conservation of crop genetic resources through maintenance of traditional farming systems. Econ. Bot. 41:86-96.

Brush, S.B. 1991. A farmer-based approach to conserving crop germplasm. Econ. Bot. 45(2):153-165.

Ceccarelli, S., S. Grando and R.H. Booth. 1996. International breeding programme and resource-poor farmers: crop improvement in difficult environments. Pp. 99-116 *in* Participatory Plant Breeding (P. Eyzaguirre and M. Iwanaga, eds). IPGRI, Rome.

Eyzaguirre, P. and M. Iwanaga. 1996. Farmers' contribution to maintaining genetic diversity in crops, and its role within the total genetic resources system. Pp. 9-18 *in* Participatory Plant Breeding (P. Eyzaguirre and M. Iwanaga, eds). IPGRI, Rome.

FAO. 1996. The state of the World's Plant Genetic Resources for Food and Agriculture. FAO, Rome.

Iglesias, C. and L.A. Hernandez R. 1996. Methodology development issues for participatory plant breeding of root and tuber crops. Pp. 129-134 *in* New Frontiers in Participatory Research and Gender Analysis. Proc. of the International Seminar on Participatory research and Gender Analysis for technology Development, Sept 9-14 1996. CGIAR Program on PRGA, CIAT.

IPGRI. 1996. An IPGRI strategy for the *in situ* conservation of agricultural biodiversity. IPGRI, Rome (unpublished).

Jarvis, D. and T. Hodgkin. 1998. Farmer decision-making and genetic diversity: linking multidisciplinary research to implementation on-farm. *In* On-farm conservation: Issues and Case Studies (S. Brush, ed.). (in press)

Jarvis, J.I., T. Hodgkin, P. Eyzaguirre, G. Ayad, B. Sthapit and L. Guarino. 1998. Farmer selection, natural selection and crop genetic diversity: the need for basic dataset Pp. 1-8. *in* Strengthening the scientific basis of *in situ* conservation of agrobiodiversity on-farm. Options for data collecting and analysis. Proc. of a workshop to develop tools and procedures for *in situ* conservation on-farm, 25-29 August 1997, Rome, Italy. IPGRI, Rome.

Joshi, A. and J.R. Witcombe. 1996. Farmer participatory crop improvement. II. Participatory varietal selection, a case study in India. Exp. Agric. 32:461-477.

Joshi, K.D., B.R. Sthapit, R.B. Gurung, M.B. Gurung and J.R. Witcombe. 1997. Machhapuchre-3 (MP3), the first rice variety developed through a participatory plant breeding approach released for mid to high altitudes of Nepal. IRRN 1997:12.

Maurya, D.M., A. Bottrall and J. Farrington. 1988. Improved livelihoods, genetic diversity and farmer participation: a strategy for rice breeding in rainfed areas of India. Exp. Agric. 24:311-320.

PRGA. 1999. Crossing perspectives. Farmers and scientists in participatory plant breeding. CGIAR Program on Participatory research and Genetic Analysis. CIAT, Cali.

Richards, P. 1996. Farmer knowledge and plant genetic resources management. Pp. 52-93 *in* In situ conservation and sustainable use of plant genetic resources for food and agriculture in developing countries (J. Engels, ed.). IPGRI, Rome, Italy.

Salazar, R. 1992. MASIPAG: alternative community rice breeding in the Philippines. Appropriate Technol. 18:20-21.

Sperling, L., M. Lovevinsohn and B. Ntabomvura. 1993. Rethinking the farmers' role in plant breeding: local bean experts and on-station selection in Rwanda. Exp. Agric. 29:509-519.

Sthapit, B.R., K.D. Joshi and K.D. Subedi. 1994. Consolidating farmers role in plant breeding: A proposal for developing cold tolerant rice varieties for the hills of Nepal. LARC Discussion paper No. 94/ Pokhara, Nepal: Lumle Agricultural Centre.

Sthapit, B.R., K.D. Joshi and J.R. Witcombe. 1996. Farmer participatory crop improvement. III. Participatory plant breeding, a case study for rice in Nepal. Exp. Agric. 32:479-496.

Weltzein/Smith, E., L.S. Meitzner and L. Sperling. 1999. Technical and Institutional Issues in Participatory Plant Breeding-Done from a Perspective of Formal Plant Breeding: A Global analysis of issues, results and current experience. Working Document No. 3, October 1999. CGIAR, Systemwide Program on PRGA for Technology Development and Institutional Innovation

Witcombe, J.R., A. Joshi, K.D. Joshi and B.R. Sthapit. 1996. Farmer participatory cultivar improvement. I: varietal selection and breeding methods and their impact on biodiversity. Exp. Agric. 32:445-460.

Witcombe, J. 1999. Do farmer-participatory methods apply more to high potential areas than to marginal ones? Outlook on Agriculture 28(1):43-49.

Witcombe, J.R. 1997. Participatory approaches to plant breeding and selection. Biotechnology and Development Monitor No 29, December 1996.

11. Experiences in participatory approaches to crop improvement in Nepal

Anil Subedi, Krishna Joshi, Pratap Shrestha and Bhuwon Sthapit

Introduction

The adoption rate of modern high-yielding crop cultivars is usually very low in most developing countries. The crop varieties bred by the national research systems in these countries, including Nepal and India, must pass through a highly centralized process of varietal testing, release and certification (Sthapit 1995; Witcombe *et al.* 1996). By the time materials reach farmers, the choice of seed for farmers is already restricted. At this stage of on-farm testing, the farmers' only option is to accept or reject a few finished cultivars. In Nepal, 43 varieties and in India, more than 500 varieties of modern rice have been released, but adoption of released varieties by the farmers of marginal environments has been poor (Maurya *et al.* 1988; Chemjong *et al.* 1995; LARC 1995; Joshi and Witcombe 1996).

Crop varieties generated through formal breeding have largely been suitable for resource-rich and high-production environments. The centralized breeding programmes focus mainly on genetic yield potential, minimizing genotype-by-environment (GxE) interactions, and seeking broad adaptation as a means to maximize production to address national food demands (Eyzaguirre and Iwanaga 1996). The crop varieties bred through this system, therefore, require optimal conditions or high inputs. Most farmers cannot afford external inputs to modify their production systems and create uniform conditions. Since farmers in developing countries live in diverse agro-ecological conditions and manage a number of complex and low-input management practices, GxE needs to be exploited, not minimized (Sthapit and Subedi 1999). This requires approaches to crop improvement that focus on varietal needs of specific niche environments and complement the knowledge and requirements of the community through adequate farmers' participation.

In conventional plant breeding, development of varieties is considered to be the task of plant breeders while farmers are supposed to wait for many years for the finished products. Farmers' breeding knowledge is less valued and the importance of their participation in the crop improvement processes is not well recognized. Modern crop varieties have usually failed, especially in diverse agro-ecological and sociocultural environments, to meet farmers' multiple goals, resulting in poor performance and low adoption of these varieties. These realities have forced scientists and development workers to rethink the whole approach to crop improvement and have encouraged many of them to experiment with innovative approaches that take care of the shortcomings mentioned above. This article shares the experiences of LI-BIRD in implementing some new approaches to crop improvement in Nepal.

Participatory approaches to crop improvement

Over the last four years, LI-BIRD has been working on a number of participatory approaches to crop improvement in its quest for ways to meet farmers' multiple varietal needs (Table 11.1). Three such approaches, namely (1) informal research and development, (2) participatory variety selection, and (3) participatory plant breeding, are discussed here. Farmers' participation is central in these approaches, and decentralized testing and participatory evaluation are inbuilt components.

Informal research and development

Informal Research and Development (IRD) is an informal and simple method of testing, choosing and multiplying seeds of choice for development (Joshi and Sthapit 1990). The main purpose of IRD is to overcome the limitation of poor access to new crop cultivars by farmers that exist in

Table 11.1. PVS and PPB programmes of LI-BIRD in different institutional settings, 1996-98

Support institutions	Types of collaboration	Nature of work
PVS (Participatory Variety Selection)		
Annapurna Conservation Area Project (ACAP)	NGO/NGO	LI-BIRD provides PVS materials and PPB products on cost basis; ACAP staff conduct PVS on-farm
CARE, Nepal	NGO/International NGO	CARE Nepal and LI-BIRD partnership PVS in parts of two districts (low to mid hill and *Terai*) on pilot basis; Capacity-building of CARE/N staff in PVS
PLAN International, Nepal	NGO/International NGO/ Resource-poor farming communities	PVS of food and vegetable crops in parts of *Terai* district of Sunsari
GARDP – II (EU-funded) implemented jointly with His Majesty's Govt. of Nepal line agencies	NGO/Village Development Committee/GARDPII/District Line Agency	PVS on food crops in one hill district
Plant Sci. Prog., DFID-supported but solely managed by LI-BIRD	NGO/farming communities/ Univ. of Wales/UK	PVS and IRD main focus for methodological test in high yield potential areas and to demonstrate increased cultivar diversity and accelerate adoption through participatory approaches
Using Diversity Research Award/IDRC-funded PVS	NGO/Community-Based Organizations	Landrace enhancement of upland (*Ghaiya*) rice in low hill parts of two hill districts
PPB (Participatory Plant Breeding)		
IPGRI/Italy (funds from The Netherlands government)	NARC/NGO/CGIAR/CBOs	Strengthening the scientific basis of *in situ* conservation of agricultural biodiversity. PPB in rice to create varietal diversity and add value to local landraces.
Plant Sci. Prog., DFID	NGO/Farming communities	PPB on rice in high production potential system of *terai* and low to mid-hill areas. Mutation breeding initiated to remove excessive awns of Pusa Basmati: Crossing of IR44595 x K-III (value addition).
System Wide Program – Participatory Research and Gender Analysis	CGIAR/NGO/NARC/CBOs	Farmer-led maize breeding, transfer of breeding skills to farmers, methodology testing, variety development, and empowerment
Sainsbury Family Trust/UK	NGO/CBOs	PPB in upland rice (*ghaiya*) in marginal areas of two hill districts
LI-BIRD	NGO/Farming Communities	Monitoring of PPB products in high-altitude villages of western hills to evaluate cost-effectiveness of PPB. Initiation of mutation breeding on M-3 to reduce shattering loss

Source: LI-BIRD Project Briefs 1998.

the conventional research and extension systems. IRD, first used at Lumle Agricultural Research Centre (LARC) in Nepal, is now increasingly being used for participatory variety testing and dissemination by different organizations in the marginal as well as high-potential environments of Nepal and India (Joshi and Sthapit 1990; Joshi et al. 1997a, 1997b, 1998)

Methodological processes

The IRD approach to crop improvement believes that farmers have knowledge and expertise related to the suitability of new crops and crop varieties in their local conditions but are constrained by their limited access to new germplasms. Box 11.1 describes the methodological processes of IRD.

Prior to implementation of IRD, a *Samuhik Bhraman*[1] is carried out with the community members to assess prevalent agroclimatic and cropping conditions, varietal diversity, and farmers' needs and preferences. This is followed by a rapid farmers' network analysis (FNA)[2] which has been included recently as a tool to identify key nodal farmers for testing and rapid diffusion of farmer-preferred varieties. The selected farmers are then given a small quantity of seed of released and/or nearly finished varieties to grow in their own management conditions. No external inputs or instructions on how to grow them are given to the experimenting farmers.

Adoption rates are monitored after several seasons to find out the most popular cultivars among farmers. This is done through field visits by the researchers followed by informal interviews with participating farmers (Joshi et al. 1997a, 1997b). Formal evaluation of the performance of IRD crop varieties (relative to existing varieties) is not made. Anecdotal records also have been used in recent years to find out the overall performance of crop varieties and the pattern of their adoption and spread in IRD villages (Joshi et al. 1998).

Effectiveness/scaling-up

IRD has been very cost-effective in variety testing and in promoting quick adoption and

---

Box 11.1. IRD processes and participatory tools used.

- *Samuhik Bhraman* (a kind of RRA) for rapid assessment of local situation, agro-ecology, production systems, existing varietal choice, and farmers' need and preference.
- *Farmers' network analysis* (FNA) to identify key/nodal farmers for testing and rapid diffusion of farmer-preferred varieties.
- Distribution of small seed kits containing releases, pre-releases and landraces.
- Monitoring of varietal adoption by sample survey after a few seasons.

(Source: Joshi and Sthapit 1990)

---

[1] *Samuhik Bhraman* is a participatory rural appraisal tool in which a team of interdisciplinary scientists travels together.
[2] A network in a social system refers to the inter-relationship among a set of individuals. Understanding and examination of such network relationships provide significant insights about social and individual behaviour, and the process by which diffusion of innovations occurs among people in a social system. Network analysis can obtain information on all relations among network members (Burt 1980). The importance of such networks has been empirically demonstrated in the farming communities of the western hills of Nepal (Subedi and Garforth 1996).

dissemination of the preferred varieties by farmers. An evaluation study on *Chaite* rice in the marginal areas of the western hills of Nepal indicated that IRD was 43% more cost-effective than the formal system of technology generation and extension (Box 11.2). A recent evaluation of participatory crop improvement programme in the high-potential production areas of Chitwan and Nawalparasi districts of Nepal revealed that estimated benefits from IRD were 33% greater than the estimated costs (DTZ Peida 1998). IRD also has been instrumental in the spread of new crop varieties through farmer-to-farmer seed exchange (Box 11.3).

Participatory Variety Selection

Participatory Variety Selection (PVS)[3] is the selection of fixed lines (including landraces) by farmers in their target environments using their own selection criteria (Joshi and Witcombe

---

Box 11.2. Preference and adoption levels of different *Chaite* rice varieties included in IRD programme in the Western Development Region of Nepal, 1991-93.

Total sets distributed	1803
Households surveyed in 1992	238 (13)
Households willing to continue	140 (59)
Households surveyed in 1993	92
Households adopting new *Chaite* rice varieties	34 (37)
Households willing to test new varieties	52 (57)

(Source: Joshi *et al.* 1997b)

Box 11.3. Spread of a preferred variety tested under IRD.

Anecdotal and direct observations are used to monitor the extent of varietal uptake of new rice varieties under the IRD programme. Farmer-to-farmer seed spread in IRD villages of East and West Chitwan created very interesting outcomes. At Chimni Tole of East Chitwan where tube well is the source of irrigation and chaite rice was never grown before, Kalinga III best fitted to the local condition when it was introduced. Mr Achyat Lamsal was the first farmer who grew Kalinga III during 1997. He multiplied it as a main-season rice crop to produce over 350 kg. Most of it was used for sowing in 1998 chaite season by the farmer while nine other farmers purchased the seed from him. Together they produced 7015 kg in one year starting from an IRD kit of 1 kg. Farmers of IRD villages were often more receptive and showed more interest in testing new varieties. Perhaps this was because the number of staff visits to IRD villages was minimal. Overall IRD trials gave similar results to that of Participatory Variety selection for the most preferred varieties. (Source: Joshi *et al.* 1998)

---

[3] PVS also encompasses participatory landraces selection (PLS) which is carried with the same process as for PVS except that all the genetic materials used in the PLS are landraces instead of modern cultivars.

1996). It is assumed that poor adoption of new cultivars is associated with inefficiencies in varietal testing and promotion rather than lack of adaptation of cultivars. Therefore, PVS is designed to identify and overcome the constraints of poor adoption rates of appropriate cultivars. Providing varietal choice and promoting participatory approaches to testing, selection and informal dissemination are central to PVS.

Methodological process

The methodological process of PVS includes four sequential steps. These steps and the tools and processes used (see Box 11.4) are described below.

Situation analysis and farmers' varietal need identification. Since adoption of a crop variety depends largely on the agro-ecological and sociocultural environments of a community, a good understanding of the local situation is the precursor to any PVS programme. A number of PRA tools are used to understand the local situation, varietal (including landraces) diversity and their associated traits. Social and resource maps of the villages are drawn with the help of a few knowledgeable farmers (key informants), both men and women, to identify ethnic and resource diversity of the villages. Wealth ranking is also done through the same group of farmers to categorize farming households in different socioeconomic groups based on their own criteria, and to identify differences among them in access to and control over production resources. Varietal needs and preferences of different categories of farmers for different crops/varieties are done through a focus group discussion (FGD) using crop history and preference matrix ranking techniques. Depending on the scope and objectives of the PVS programme, a household baseline survey, using formal sample survey techniques, is also done to establish baseline information useful in monitoring impacts of the programme.

Search for suitable genetic materials. Following the situation analysis and need identification, suitable crop varieties or landraces are sought from different sources within or outside the country on the basis of farmers' preferences for particular trait(s). Seeds of new crop varieties collected are then packed into small packets of 0.25 to 2.0 kg, depending on the size of the seeds, to cover land areas of less than 0.05 ha. It is important to keep to a minimum to minimize the risk of crop failure.

Farmers' experimentation of new crop varieties in their own field and management practices. Based on the information on rapid farmers' network analysis, seeds of different crop varieties procured from the search process are distributed to the interested nodal farmers in the target communities for experimentation. One farmer is given only one crop variety but the same variety is given to more than six farmers belonging to different wealth categories and ethnic groups. This facilitates evaluation of the given crop varieties across different categories of the farmers in the community. The testing is done on farmers' own farms and under their own management practices without intervention from the researchers. This enables farmers to carry out their own research and assess the suitability of new crop varieties for their land and other farm resources (inputs), and their consumption as well as other needs, such as marketing, processing, livestock feeds and so on.

A joint farm walk by a group of participating and non-participating farmers, breeders, researchers, social scientists and other development workers is done at the crop maturity stage in order to observe and assess the performance of crops in farmers' fields. During the field visit, characteristics of each crop variety are assessed and comparisons are made between and among the varieties. The farm walk is followed by FGD to further identify positive and negative traits as perceived by the farmers. Preference ranking of each variety is then done between tested varieties by mutual consensus among the farmers. This tool is useful to understand how farmers trade off yield potential of cultivars for certain important varietal characteristics.

About 3-4 months after the harvest of the crops, a **post-harvest evaluation** of the PVS crop varieties is done through a FGD of the participating farmers. Facilitated by researcher(s), farmers are encouraged to assess the crop varieties by taking into account all relevant qualitative traits. Experiences have shown that, in addition to yield, a farmer's decision to reject or adopt a crop variety is also based on its post-harvest qualities such as threshing quality, milling recovery, straw quality, and a number of culinary qualities such as cooking quality, palatability, taste and so on. Since women farmers play a main role in assessing such traits and make final decisions on rejection or adoption of a variety, their participation in the process is crucial. Alternatively, detailed feedback from the participating farmers is also collected for statistical evaluation of PVS crop varieties against pre- and post-harvest traits across agro-ecological variations, and farmers' wealth categories and ethnicity as well as to understand the adoption behaviours of the farmers in the community. A short household questionnaire is used for this purpose.

Wider dissemination of farmer-preferred crop varieties. The farmers testing new crop varieties keep seed of preferred varieties for the next season and usually expand the area for cropping. The successful varieties also find their way to other farmers within and outside the community through farmer-to-farmer seed exchange. Farmers are also encouraged and facilitated to organize into groups to produce seeds of preferred varieties locally and to sell to interested farmers within and outside the community. Such a decentralized seed production and distribution system has been successful in the LI-BIRD and CARE Nepal collaborative PVS programme.

Effectiveness/scaling-up

PVS has been effective in identifying farmer-preferred varieties from a wide range of choices in a short period of time. The major indicators of adoption by farmers are seed saving by participant farmers for re-sowing, local seed supply among the farmers through their networks or purchase

---

Box 11.4. PVS process and participatory tools used.

Steps of PVS	Participatory tools used
1. Situation analysis and identification of farmers' varietal needs	• RRA/PRA; Samuhik Bhraman • Preference matrix ranking
2. Search for suitable genetic materials.	• Computer database • Indigenous knowledge system
3. Farmers' experimentation of new crop varieties in own fields and with own management practices.	• Social and resource mapping • Wealth ranking • Site characterization using transect • Farmer network analysis • Farm walk • Focus group discussion • Preference matrix ranking
4. Wider dissemination of farmer-preferred crop varieties.	• Participatory monitoring of varietal spread • Ball and string technique

(Source: Adapted from Joshi and Witcombe 1996).

of seeds of new varieties from the resource organizations. In a study of 80 households participating in PVS trials on *Chaite* rice, within one year of intervention 49% of them saved sufficient seeds for re-sowing in 20 ha. An additional 17% of farmers purchased seeds from their neighbours (Joshi et al. 1998). In a similar study, the varietal uptake pattern for main-season rice varieties also was quite high. Following a single-season exposure to new varieties through PVS, out of 250 farmers, 121 (48%) saved seeds, and 27 (11%) demanded seed for new main-season rice varieties. Seed retention by farmers and farmer-to-farmer seed exchange or sale is expected to cover 35 ha of rice land. Since these are preliminary findings, there may be variation in adoption rate over time.

Participatory Plant Breeding

Participatory Plant Breeding (PPB) is a breeding process in which farmers and plant breeders jointly select cultivars from segregating materials under a target environment (Sthapit et al. 1996; Witcombe et al. 1996). Other forms of PPB may also include activities such as germplasm enhancement through pure line or mass selection. PPB is generally done when the possibilities of PVS have been exhausted, or when the search process in PVS has failed to identify any suitable cultivar for testing or if farmers identify a new problem in the existing cultivars. It is desired that at least one of the parents should be a local landrace screened under target habitats utilizing farmer selection criteria and knowledge. PVS and PPB are cyclic processes and interlinked to each other (Fig. 11.1), but they have distinct processes as well.

In PPB, farmers are involved at much earlier stages in the breeding process to identify farmer-acceptable cultivars more efficiently. It also strengthens on-farm conservation and selection of locally adapted and farmer-preferred genetic materials through skill transfer and empowerment. PPB could be either consultative or collaborative, based on the typology of farmer participation suggested by Biggs (1989). In consultative PPB, farmers are consulted to set breeding goals and choose appropriate parents and testing sites. In collaborative PPB, in addition to setting breeding goals, farmers grow segregating genetic materials and select the best plants among them in their own fields. Subedi et al. (1997), however, argued that it is difficult to standardize the degree of farmer participation as the skills and strengths of participant farmers as well as institutional settings undertaking PPB may greatly differ. Table 11.2 shows various types of PPB based on level of participation. The success of PPB, therefore, rests on blending the comparative strengths of farmers, breeders and social scientists involved in the process. The use of consultative or collaborative methods as well as choosing the right level of farmer participation depends on the crops, the capacity of participant farmers, willingness and availability of breeder and researcher resources.

Methodological process

The methodological processes involved in PPB can differ from case to case depending on the nature of the problem, institutional settings and the degree of farmers' participation achievable in a sociocultural setting. The processes of the PPB that involve farmers from the very beginning of the breeding activities are discussed here (Box 11.5).

Need identification and setting breeding goal. The first and foremost step in PPB is the identification of farmers' problems, needs, preferences and priorities with the farming community. The situation vis-à-vis genetic resources, their shortcomings and scope for their improvement is understood through situation analysis of the area, focus group discussion and other PRA tools. Once the situation is clearly understood, the next step is to set the breeding goal in a meeting with the community members. The goal may be for yield improvement, early maturity, quality or any other traits that are technically feasible.

Local PGR management and participatory crop improvement

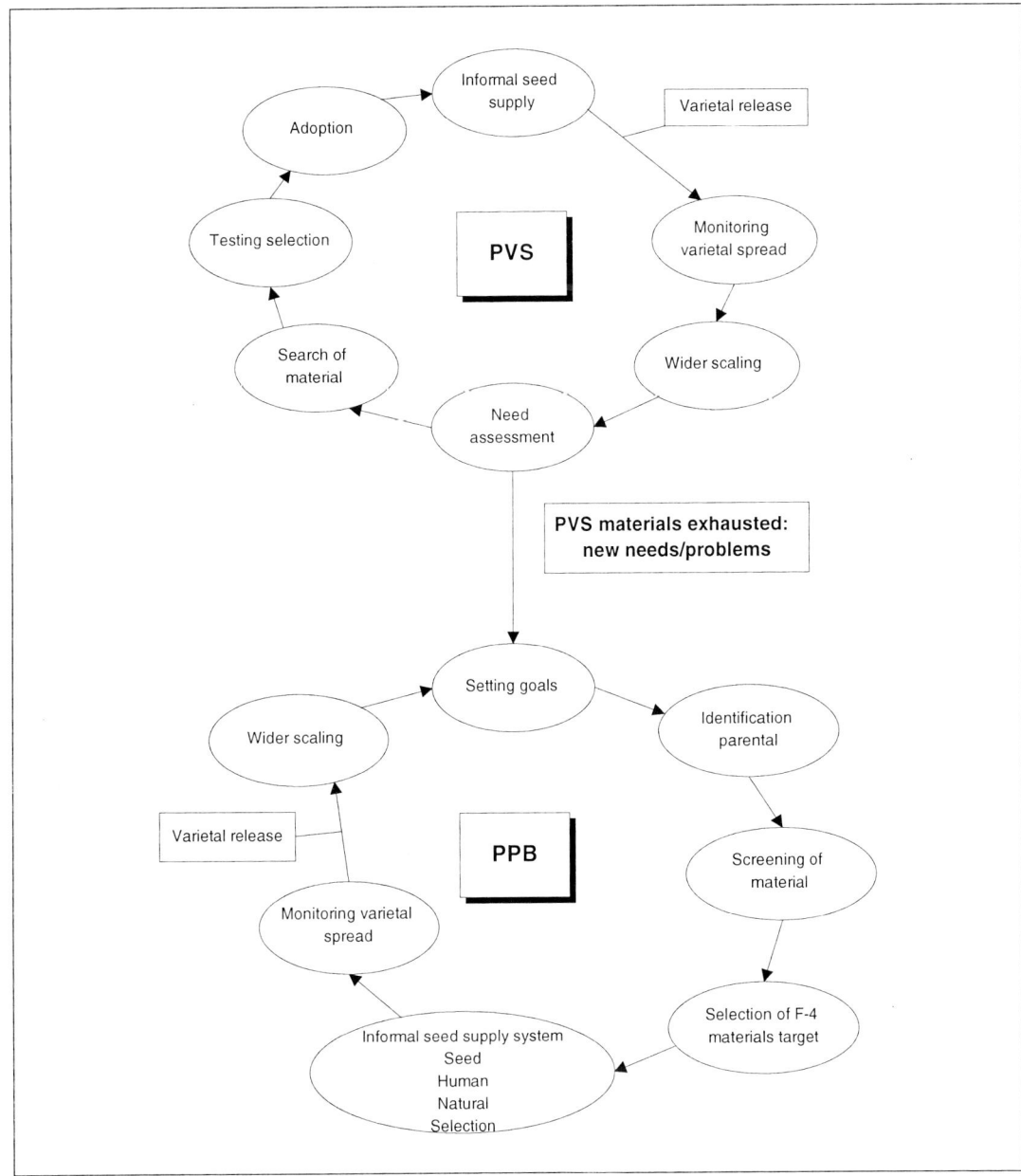

Fig. 11.1. PVS and PPB cycle: the process of crop improvement and generating genetic resources base.

Parent selection and generating new variability. In PPB, at least one parent is selected from the local landraces to ensure local adaptation and to incorporate other useful traits. A cultivar identified from the PVS programme or a well-preferred modern cultivar also can be used as a parent. As farming communities know the strength and weaknesses of their local varieties, parent selection is planned with the community. This is done through preference matrix ranking of varieties of the crop in question. The selection of the donor parent is done using breeders' expert knowledge. In self-pollinated and vegetatively propagated crops, creating new variability needs expert skills. There are many examples where farmers can also create genetic variability among cross-pollinated crops.

Expert farmer selection and role identification. Since the involvement of all farmers in a community in specific PPB tasks is not possible, a few interested, research-minded, knowledgeable (about crops used for PPB) farmers are involved in the PPB work. Selection of such experts, who also occupy nodal positions in the community networks, is done through rapid FNA and group discussion with the farming community. Farmer's knowledge and skills are valuable in screening segregating materials for abiotic variations, whereas screening of segregating materials for biotic resistance can be done jointly or by a breeder/pathologist.

Selection of segregating lines. Farmers have the skills and knowledge to select crop varieties on the basis of their phenotypic characteristics. Improving farmers' ability to select for simple traits or other desirable traits requires transfer of knowledge of scientific methods of seed selection from plant breeders to farming communities. This is done jointly by breeders and farmers during the Farm Walk and Focus Group Discussion directly in the field. Farmers select the individual plants based on the criteria that they consider important for their conditions and preferences. Breeders facilitate them in further selection from the segregating lines by pointing out the traits that are important but overlooked. The seeds of the selected lines are planted in the following year in the same environment and further selections are made to stabilize the population.

Effectiveness/scaling-up

A monitoring study in the western hills of Nepal showed that within 2 years, 33% of the households covering up to 50% of the rice area were covered by the rice varieties developed through PPB (Sthapit *et al.* 1998). Farmers' informal seed-supply systems are the major outlets for the scaling-up of PPB products. Farmers' informal network of relations is the major channel for the spread of the varieties within or outside the community followed by the adoption by neighbourhood farmers (Sthapit *et al.* 1998). However, the diffusion process has to be accelerated through other means of scaling-up such as CBOs, private entrepreneurs, NGOs as well as public extension networks.

The Nepal case demonstrated that the products of PPB also can be entered into the formal multilocational trials and that varieties selected by farmer-breeders can be officially released (Joshi *et al.* 1997a). The breeder can pick up the most widely accepted materials from PPB and introduce them into a varietal testing system to test wide adaptation, yield performance and other traits. Variety release and formal seed production are still the desirable end-product to make the results of the PPB widely available, but it is first necessary to simplify testing procedures, variety release and the seed regulatory framework to accommodate PPB products. Issues of intellectual property rights have been raised by outsiders. The IPGRI-supported project in Nepal used PPB as a strategy for biodiversity enhancement and production which could be achieved if PPB products are spread through informal "farmer-to farmer" seed-supply systems.

Contribution of participatory crop improvement approaches to genetic resources conservation and use

Germplasm enhancement. It is one of the important methods of improving crop germplasm and can be effective even with the community. It is established that about 15% yield advantage can be experienced with selection and population improvement. There are a number of examples of crop varieties developed through germplasm enhancement. Chhomrong *Dhan* released for high-altitude areas of Nepal was developed through pure line selection (Sthapit 1995). Similarly Lumle soyabean, Tikot fieldpeas and Lumle *Tori* are varieties developed through this process

Table 11.2. Methods of plant breeding in self-pollinated crops with varying degrees of farmer participation

Methods in increasing order of farmer participation	Site specificity	Example
1. All generations grown by plant breeders on-station. Farmers involved at pre-release stage or even after release	Wide adaptation. Early generations may all be in a single location followed by multilocational testing.	CGIAR, NARS
2. Early generation (F_2) in farmers' fields. All other generations and procedures with plant breeder on-station	Single location testing for F_2.	Thakur (1995)
3. Best advanced lines at F_7 or F_8 given to farmers for testing. Closest method to PVS since farmers given nearly finished products.	Easy to test best advanced lines across locations.	Galt (1989); Joshi and Sthapit (1990); Joshi and Witcombe (1996); Joshi et al. (1997a)
4. From F_2 or F_3 onwards farmers and plant breeders collaborate to select and identify the best material on-farm. Farmers select. Plant breeders facilitate the process. Release proposal prepared by plant breeder. Farmer-breeder name included.	Possible to run selection procedures on early generations. Parallel testing of promising lines in formal breeding programme.	Sthapit et al. (1996): F_5 generation lines were used in this case.
5. Breeder gives F_3 or F_5 bulk to farmers. All selection left to farmers. At F_7 to F_8 stage or later, breeders monitor diversity in farmers' fields and identify best material to enter in conventional trials.	Extremely easy to run selection schemes in many locations.	Salazar (1992)
6. Trained expert farmers make crosses and do all selection with or without assistance from breeders.	Specific to farmers' requirements.	Savaliya 1997; Seta 1997

(Source: Witcombe et al. 1996 with some additional information)

(Joshi et al. 1997a, 1997b). There is great scope for the involvement of communities in germplasm enhancement; for example local varieties can be improved through the PVS approach wherein the role of farmers will be to select the desirable plant types from a variable population of a landrace.

In addition to germplasm enhancement and easy access to these materials by rural communities, implementation of participatory approaches to crop improvement in Nepal has been shown to have the following impacts.

Local/specific adaptation of new varieties generated. Selection under local conditions generates materials that are stress tolerant, have adaptive traits and are well liked by farmers.

On-farm conservation of landraces. The breeding strategy that employs crossing of landraces with modern cultivars adds value to the landraces. This makes these landraces more attractive to farmers for continued cultivation (Sthapit et al. 1996).

Empowerment to farmers. The participating farmers are empowered through their involvement in defining breeding goals and selection of the genetic materials as well as through transfer of skills from the scientists. They will also have direct control over improved seed.

> **Box 11.5. PPB process and participatory tools used.**
>
> | Process | Participatory tools used |
> |---|---|
> | **1. Need identification and setting breeding goals:** | |
> | • understanding reasons for growing diverse varieties | RRA/PRA; village workshop, FGD, preference matrix ranking; diversity fair, community biodiversity register (CBR) |
> | • setting breeding goals and roles jointly to meet immediate need | Focus Group Discussion (FGD) |
> | **2. Parent selection and generating new diversity:** | |
> | • identification and use of locally adapted varieties as parent materials | Diversity block, PRA, FGD |
> | **3. Expert farmer selection and role identification:** | |
> | • identification and selection of knowledgeable farmers having interest in PPB work | Farmers' Network Analysis (FNA) Joint farm walk to exchange breeding knowledge between breeders and farmers. |
> | • farmers assume role of selecting suitable cultivars from the segregating materials | Diversity fairs |
> | **4. Selection of segregating lines:** | |
> | • decentralized selection of segregating lines (variable population) by farmer under target environment | Farm walk Farm school (for selection training on site) FGD Preference ranking |
> | • post-harvest evaluation using gender perspective | Preference matrix ranking |
> | **5. Variety release and distribution:** | |
> | • varietal spread through informal seed supply system | ball and string technique |
> | • release variety on basis of varietal spread data | PRA monitoring |
>
> (Source: Updated and modified from Sthapit and Subedi 1999)

Cultural and equity considerations. PPB addresses the needs of farming communities living in the most marginal environments. These people largely represent ethnic minorities, and the use of crops and/or crop varieties often relates to their sociocultural identity. PPB provides scope for considering such niche-specific needs and priority in the variety development process.

Acknowledgements

The PPB methodology reported herein would not have been possible without the sincere collaboration of farmers from Chhomrong and Ghandruk villages of Western Hills of Nepal.

References

Biggs, S.D. 1989. Resource-poor farmer participation in research: a synthesis of experiences from nine national agricultural research systems. OFCOR-comparative study paper no. 3. Special series on the organisation and management of on-farm client oriented research (OFCOR). International Service for National Agricultural Research (ISNAR), The Hague.

Burt, R.S. 1980. Models of network structure. Ann. Rev. Soc. 6:79-141.

Chemjong, P.B., B.H. Baral, K.C. Thakuri, P.R. Neupane, R.K. Neupane and M.P. Upadhaya. 1995. The impact of Pakhribas Agricultural Centre in the Eastern Hills of Nepal: farmer adoption of nine agricultural technologies. Pakhribas Agricultiural Centre, Dhankuta, Nepal.

DTZ Peida. 1998. An evaluation study of participatory crop improvement in Nepal. A final Report. DTZ Peida Consulting, Edinburgh, UK.

Eyzaguirre, P. and M. Iwanaga. 1996. Farmers' contribution to maintaining genetic diversity in crops, and its role within the total genetic resources system. Pp. 9-18. *in* Participatory Plant Breeding (P. Eyzaguirre and M. Iwanaga, eds). Proceedings of a workshop on participatory plant breeding, 26-29 July 1995, Wageningen, The Netherlands. IDRC /FAO /CGN/ IPGRI, IPGRI, Rome, Italy.

Galt, D. 1989. Joining FSR to Commodity Programme Breeding Efforts earlier: Increasing Plant Breeding Efficiency in Nepal. Agricultural Administration (research and extension) Network: Network Paper 8. Overseas Development Institute, London.

Joshi, A. and J.R. Witcombe. 1996. Farmer participatory cultivar improvement. II: participatory varietal selection in India. Exp. Agric. 32:461-477.

Joshi, K.D. and B.R. Sthapit. 1990. Informal Research and Development (IRD): A new approach to research and extension. LARC Discussion Paper 1990/4. Lumle Agricultural Research Centre, Pokhara, Nepal.

Joshi, K.D., B.R. Sthapit, R.B. Gurung, M.B. Gurung and J.R. Witcombe. 1997a. Machhapuchre-3 (MP3), the first rice variety developed through a participatory plant breeding approach released for mid to high altitudes of Nepal. IRRN 1997:12

Joshi, K.D., M. Subedi, R.B. Rana, K.B. Kadayat and B.R. Sthapit. 1997b. Enhancing on-farm varietal diversity through participatory varietal selection: a case study for *Chaite* rice in Nepal. Exp. Agric. 33:335-344.

Joshi, K.D., R.B. Rana, B. Gadal and J.R. Witcombe. 1998. The success of participatory varietal selection for Chaite rice in high potential production systems of in the Nepal Terai. *In* Proceedings of International conference on Food Security and Crop Science, Hissar, India, 3-6 November 1998. Hisar Agricultural University, Hisar, India (in press).

LARC. 1995. The adoption and diffusion and incremental benefits of fifteen technologies for crops, horticulture, livestock and forestry in the Western Hills of Nepal. LARC Occasional Paper 95/1. Lumle Agricultural Research Centre, Pokhara, Nepal. pp 69.

Maurya, D.M., A. Bottarall and J. Farrington. 1988. Improved livelihoods, genetic diversity and farmer participation: a strategy for rice breeding in rainfed areas of India. Exp. Agric. 24:311-320.

Salazar, R. 1992. MASIPAG: alternative community rice breeding in the Philippines. Appropriate Technology 18:20-21.

Salvaliya, T.B. 1997. New variety of groundnut. *In* International Conference on Creativity and Innovation at grassroots for Sustainable Natural Resource Management, 11-14 January 1997. Centre for Management in Agriculture, Indian Institute of Management, Ahmedabad, India.

Seta, R.B. 1997. Groundnut breeder. *In* International Conference on Creativity and Innovation at Grassroots for Sustainable Natural Resource Management, 11-14 January 1997. Centre for Management in Agriculture, Indian Institute of Management, Ahmedabad, India.

Sthapit, B.R. 1995. Variety testing, selection, and release system for rice and wheat crops in Nepal. Seed Regulatory Frame Works: Nepal. ODI/UK, CAZS/Lumle Agricultural Research Centre, Pokhara, Nepal.

Sthapit, B.R. and A. Subedi. 1999. Participatory variety selection and participatory plant breeding: An NGO experience and insights to support the local genetic resource base. *In* Encouraging Diversity. The Synthesis of Crop Conservation and Improvement (W.S. de Boef and C.J.M. Almekinders, eds.). IT Publications, London

Sthapit, B.R., K.D. Joshi and J.R. Witcombe. 1996. Farmer participatory crop improvement. III. participatory plant breeding, a case study for rice in Nepal. Exp. Agric. 32:479-496.

Sthapit, B.R., K.D. Joshi, R.B. Rana and A. Subedi. 1998. Spread of varieties from participatory plant breeding in high altitude villages of Nepal. Local Initiatives for Biodiversity, Research and Development, Kaski, Pokhara, Nepal.

Subedi, A. and C. Garforth. 1996. Gender, information and communication network: implication for extension. Eur. J. Agric. Educ. & Extension 32(2):63-74.

Subedi, A., R.B. Rana and K.D. Joshi. 1997. Methodological approach to PPB: Experience from Nepal. Pp. 21 *in* Strengthening the scientific basis of *in situ* conservation of agricultural biodiversity on-farm: Options for data collecting and analysis (D. Jarvis and T. Hodgkin, eds.). Proceedings of a workshop to develop tools and procedures for *in situ* conservation on-farm, 25-29 August 1997, Rome, Italy.

Thakur, R. 1995. Prioritization and development of breeding strategies for rainfed lowlands: a critical appraisal. *In* Proc. of the IRRI conference "Fragile Lives in Fragile Ecosystems" IRRI, Los Baños, Philippines.

Witcombe, J.R., A. Joshi, K.D. Joshi and B.R. Sthapit. 1996. Farmer participatory cultivar improvement. I: Varietal selection and breeding methods and their impact on biodiversity. Exp. Agric. 32:445-460.

12. Crop improvement at community level in Vietnam

Nguyen Ngoc De

Introduction

Crop improvement has been one of the strong and continuous programmes in the Mekong Delta for major crops, especially rice and beans. However, most breeding programmes have been established and designed by breeders, neglecting the role of users: farmers and farming communities. Breeders have set their own breeding objectives and conducted crop improvement programmes based on their own analysis of problems and on-station research findings (COWI 1999). At the end of their breeding programmes, promising breeding materials were released to farmers as so-called "technology transfer". Farmers were the passive users, receiving finished varieties for their production. In many cases, farmers, especially the poor, refused to try because they did not want to take risks with new crop varieties. Resource-rich farmers were the first to try such varieties. Participation was limited to providing a piece of land to the breeders for on-farm trials. The dissemination process of "technology transfer" was very slow and costly for both breeders and farmers. As a result, the adoption of the recommended varieties in many cases has been very slow, doubtful and has even failed. Local adoption of new technologies was dependent not only on technical suitability and economical viability but also on social acceptance. The use of participatory approaches to crop improvement has assured the involvement of farmers in the whole process, or at least in the evaluation process. This has resulted in a better understanding and acceptability of new crop varieties generated through the breeding programme.

CanTho University, as the leading research institution for adopting participatory approach in rice improvement, started on-farm breeding programmes as early as after the war in 1975 by sending out their staff and students to work closely with farmers on crop improvement programmes (Xuan 1993). In 1994, with the inception of the Community Biodiversity Development and Conservation (CBDC) project, Participatory Plant Breeding (PPB) and Participatory Variety Selection (PVS) approaches were introduced as participatory methods to develop and identify suitable crop varieties specific to the niche environments and farmers' preferences (CBDC 1996, 1997).

Witcombe *et al.* (1996) and Sthapit *et al.* (1996) defined PPB as involving farmers in selecting genotypes from genetically variable, segregating materials and PVS as involving the selection by farmers of non-segregating, characterized products from plant breeding programmes. However, they also agreed that PPB is a logical extension of PVS. In our view, PVS is only the lower level of PPB. PPB, therefore, should be understood in a broader sense, as should its implication. PPB is the involvement of farmers in the whole process of plant breeding, not only the selection of segregating and non-segregating materials. Farmers can be involved in the very beginning of setting objectives for plant breeding, identifying parents, making crosses (of course with training by formal sectors), and selection of both segregating and non-segregating materials. The experiences from the CBDC project in South-East Asia have proven that point, especially in the Mekong Delta (Vietnam) and Bohol (Philippines) for rice (CBDC 1998).

These participatory approaches are also being used in one of the study sites (Tra Cu) of the global *in situ* conservation project of Vietnam, implemented in collaboration with IPGRI.

Methods used in participatory crop improvement

The participatory crop improvement programme uses participatory varietal selection (PVS) and/or participatory plant breeding (PPB) depending on farmers' varietal needs, and their breeding knowledge and technical skills. The PVS approach has been used to improve local landraces and to evaluate the finished breeding materials, obtained from research institutions, on farmers' fields. When varietal options available to farmers through PVS are limited or exhausted, PPB is initiated (CBDC 1998). Farmers with knowledge and interest in breeding are involved in PPB activities, i.e. activities from crossing of desired parent lines to selection and evaluation of the segregating genetic materials (De and Tin 1998). A flow diagram showing methods used in participatory crop improvement is presented in Figure 12.1. The methods used in implementing PPB and PVS are discussed below.

Methods used for PPB
Participatory plant breeding involves the following steps and activities.

Needs assessment and selection of cooperating farmers
Community meetings are organized to identify farmers' problems and needs to come up with suitable crop improvement strategies and plans. A group of farmers (Group 1), having knowledge and interest in breeding, are selected as cooperating farmers in consultation with the community. The breeding activities are then formulated and decided with these cooperating farmers.

Setting breeding objectives and identifying donor parents
Breeders work closely with farmers to agree on breeding objectives. Farmers have been found to use both quantitative and qualitative criteria to determine these breeding objectives. Some of the examples of such criteria are high yield, short duration, resistance to major pests and diseases, and stickiness in cooked rice. Breeders then assist farmers to search for suitable donor parents for crossing, according to the breeding objectives. These donors could be found in the available genetic materials at local level or from research institutions and are made available to cooperating farmers.

Making crosses and selecting from segregating materials
The Group 1 farmers are given additional training on crossing techniques and assisted in making desired crosses. In other cases, breeders provide seeds of segregating lines at very early generations (F_2, F_3 and F_4) to the farmers for selection of desired lines based on their own criteria. Farmers have been found to handle segregating materials as early as the F_2 generation. In the process, farmers apply their own crop-management practices. Farmers observe, evaluate and harvest the selected plants individually, according to the breeding objectives. This process is repeated until stable lines are obtained. For management reasons, the number of individual plants selected each season is limited depending on farmers' capacity for seed handling and the limited land assigned as breeding plots. Therefore, genetic variation of farmers' selection is usually narrow. Group 1 farmers are only involved in the selection process; field operations are done with the help of other farmers in the community.

Observation test
Pure lines selected from the segregating materials are planted in observation tests for adaptation and yield in which common local varieties are used as local checks. Farmers compare the performance of new lines with the local check and select promising ones for further evaluation in yield trials by Group 2 farmers.

Monitoring
The Group 1 cooperating farmers make close field observations with technical assistance from breeders and agricultural extensionists. These farmers also keep records on the field conditions and crop performance for later analysis in determining the suitability of the new crop varieties under selection.

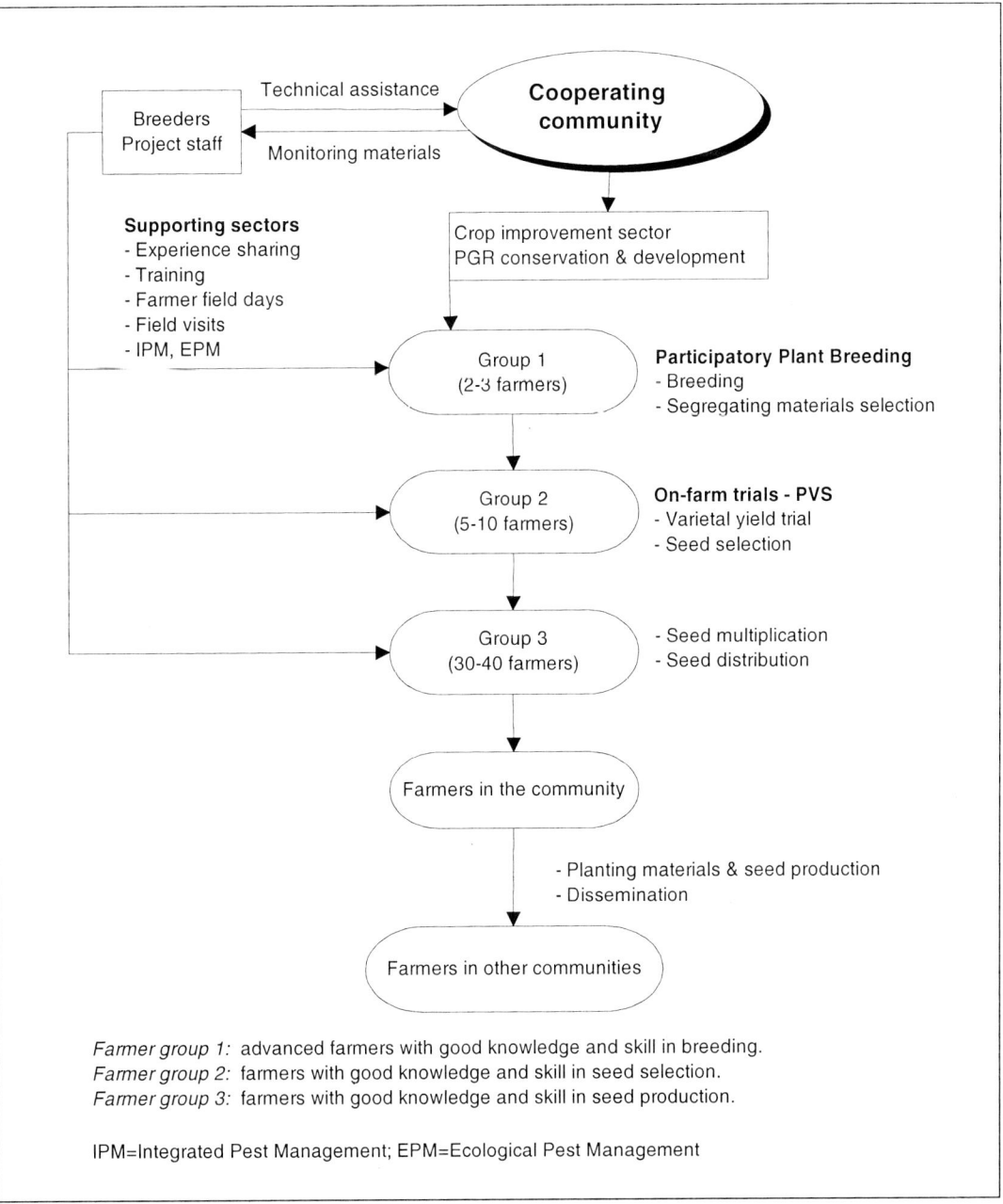

Fig. 12.1. Community-based networking diagram for PPB and PVS.

Methods used for PVS
Participatory varietal selection involves the following steps and activities.

Needs assessment and selection of cooperating farmers
As in PPB, community meetings are organized to identify farmers' problems and needs in relation to their current crop varieties (Fig. 12.1). Farmers may wish to improve their current

varieties or change to new promising varieties. A separate group of farmers (Group 2), with good knowledge and skills in seed selection and management, are selected as cooperating farmers in consultation with the community. The PVS activities are then formulated and decided with these cooperating farmers of both Groups 1 and 2.

Provision of genetic materials and participatory selection
Three sources of genetic materials are used to obtain seeds for participatory selection of desired crop varieties.

PVS with improved local landraces
The improvement of local landraces is done through mass as well as pure line selection. Since the mass selection method does not require highly specialized skills, Group 2 farmers, after simple orientation, have been able to undertake this selection. On the other hand, the pure line selection method of crop improvement requires specialized skills and care on the part of the farmers. For this reason, only Group 1 farmers have been used to do pure line selection, with adequate training and intensive monitoring. The improved local landraces are then given to a large number of farmers within the community, as PVS materials, for their own testing and selection.

PVS with re-introduced local landraces
PVS also re-introduces landraces from genebanks back to a community when its local materials have been totally destroyed by disaster. More often the collected local varieties from different locations within and outside the community are evaluated in the community to give farmers more choices among landraces.

PVS with modern crop varieties
Modern crop varieties from research institutions and finished products from PPB are also given to the cooperating farmers to test their suitability under their own management conditions and household requirements.

Yield trial of successful PVS varieties
The crop varieties preferred by the farmers under the PVS programme are then put into varietal yield trials in the community for farmers to directly observe and make selection of their choices. Common varieties in the community are used as the local check in these trials. Farmer field days are organized just before harvesting to bring farmers in the community to the trial plots for a joint evaluation of the tested varieties. Desirable varieties (usually 2-3 varieties) are then selected for seed multiplication.

Seed multiplication
Desirable varieties selected by farmers from yield trials are distributed to a group of farmers (Group 3), with considerable knowledge and interest in seed production, to multiply large quantities of seeds for use by other farmers in the community. Seed multiplication fields are closely monitored and used as a final check for large-scale production.

Monitoring
Field visits and farmer field days are the most appropriate tools for participatory monitoring and evaluation of PVS activities. Breeders, field staff, extension workers and farmers (Fig. 12.2) participate in such monitoring and evaluation activities. Data collecting depends on farmers' objectives, and includes some common traits such as growth duration, plant height, tillering capacity, grain yield and quality, and tolerance to insects and diseases.

Fig. 12.2. Thai minority farmer practising shifting cultivation of upland rice, Vietnam.

Field experiences with rice

Participatory Varietal Selection (PVS)

Rice is the major food crop in the Mekong Delta. The PVS activities on rice were undertaken in the Mekong Delta as early as the 1970s in different forms. The most common of these was varietal yield trials. The main objectives of the varietal yield trials were to generate farmer-preferred crop varieties and faster dissemination of these varieties. Can Tho University had been a leading research institution in initiating and implementing these on-farm research activities. In the beginning, breeders and researchers cooperated with advanced farmers individually throughout the Mekong Delta (De 1997).

During the period 1975-95, hundreds of promising rice varieties were tested in farmers' fields, and a number of varieties were identified and released. Some of these rice varieties are IR36 (later named NN3A), HT6 (NN6A), MTL30 (NN7A), HT19 (NN2B), IR42 (NN4B), MTL58 IR13240-108-2-2-3) and MTL87 (IR50404-57-2-2-3). These varieties have been a great contribution to the improvement of rice production in the Mekong Delta. Many farmers, such as Mr Hai Huu (Long An province), Mr Hai Chung, Mr Tu Tai, Mr Ba Chuong (Tien Giang province), Mr Ba Cung (An Giang province), Mr Muoi Tuoc, Mr Muoi Than Nong (Vinh Long province), and a few others, were known as **"rice selection kings"**. Farmers were also found to use pure line selection to improve the formally released varieties for grain quality and adaptation to specific conditions in their areas. This process in fact has strengthened on-farm conservation of crop diversity.

Since 1994, with the inception of the Community Biodiversity Development and Conservation project, PPB and PVS approaches have been included in their current form in the crop improvement programme. In these approaches a shift from dealing with the advanced individual farmers to farmer groups and farming communities has been introduced (CBDC 1998). As a result, more farmers have been involved; the degree of participation has improved and better works have been organized at grassroots level by communities themselves with help from many local authorities. Four farming communities used as pioneer organizations are Nhut Ninh community (Tan Tru district, Long An province), My Thanh community (Ba Tri district, Ben Tre province), Ke Sach community (Ke Sach district, Soc Trang province) and Long Thanh community (Vinh Loi district, Bac Lieu province). Results of PVS activities are presented in Tables 12.1 and 12.2.

Table 12.1. Number of rice varieties tested and selected from PVS activities at four communities in the Mekong Delta (CBDC 1998)

Year	Type[†]	Nhut Ninh TE[‡]	Nhut Ninh SE[‡]	My Thanh TE	My Thanh SE	Ke Sach TE	Ke Sach SE	Long Thanh TE	Long Thanh SE
1994	TR	252	8						
	DWR	20	6						
	MR	18	4						
	HYV	5	1			5	1	22	2
1995	TR	23	3						
	HYV	1	1	5	4	5	3	169	16
1996	TR		1						
	MR			22	1				
	HYV	9	9	34	9	89	?	9	1
1997	TR	222	2						
	MR	7	Loss	32	29			25	?
	HYV			20	9	16	8	20	3
1998	MR	11						12	
	HYV	12	6	18	8	19	9	24	5

[†] TR= Traditional rice; DWR= Deep water rice; MR= Medium rice; HYV= High-yielding rice (Early).
[‡] TE = Tested; SE = Selected.

Table 12.2. Common varieties selected from the PVS activities at four communities in Mekong Delta (CBDC 1998)

Rice varieties[†]	Nhut Ninh	My Thanh	Ke Sach	Long Thanh
TR	Nep Thom, Tai Nguyen, Me Huong	–	Tai Nguyen	–
MR	–	MTL83, MTL124	–	MTL83
HYV	IR49517, IR64, MTL156, 157, MTL159, 199	IR54883, S976B, MTL138, 205	MTL99, 101, MTL142, 157, MTL164, 190, MTL199, 201, MTL202	IR64, MTL138, MTL142, 147, MTL149, 150, MTL156, 157, MTL159, 199

[†] TR= Traditional rice; DWR= Deep water rice; MR= Medium rice; HYV= High-yielding rice (Early).

Participatory Plant Breeding (PPB)

In the 1996-97 dry season, the project decided to start providing segregating breeding materials from 63 F_2 populations of 12 crosses made by the Rice Research Department of Can Tho University for farmer selection in the four communities that were involved in the PVS programme (Table

12.3). These crosses are L245, L246, L247, L248, L249, L250, L251, L252, L253, L254, L255, and L256. Many farmers were interested in selecting individual plants from segregating populations by their own criteria and under their own management conditions. Some of the farmer-selected varieties are now stable lines and are being tested under yield trials.

The two promising farmer selections (L246-7-3-B and L247-1-5-B), noted by farmers as SiC-1 (Soc Trang Selection, no. 1) and SiC-2 (Soc Trsng Selection, no. 2), respectively, were purified by bulk selection method after F_4. Farmers in Ke Sach community (Soc Trang province) are now multiplying it for distribution amongst themselves. Mr Canh is the leader of this farmers' group who has led the selection activities in this community. Similarly, L246-10-1-B, a promising line selected by farmers in My Thanh community (Ba Tri district, Ben Tre province), is also under yield testing and seed multiplication.

Besides the initially selected four communities, the PPB and PVS has expanded to other individual advanced farmers in the Mekong Delta. Mr Hai Triem from An Giang province was well known as "**farmer of the era**" and awarded the "**3rd Labour Medal**" by Central Government for his contribution to rice improvement in the area.

Table 12.3. Number of segregating populations distributed and selected by four communities from PPB activities in the Mekong Delta by year

Community	No. of populations selected by generation				Farmers' selection
	F_2	F_3	F_4	F_5	
Nhut Ninh	13	13			
My Thanh	20	8	3	1	L246-10-1-B
Ke Sach	10	4	2	1	L246-7-3-B (SiC-1)
					L247-1-5-B (SiC-2)
Long Thanh	20	11			
Total	63	36	5	2	

Problems and lessons

Problems
- Low education level of farmers means they require more training and adoption of PPB is slow.
- Few farmers are interested in working with breeding and selection of segregating materials. Farmers are more willing to multiply the promising varieties than to select from segregating materials or make crosses.
- The number of collaborating farmers on PPB is limited, especially in the pedigree selection method and segregating material selection because the work is time consuming.
- Agriculture policy is more favourable to commercial production than to diversity.
- Owing to fast turnover of rice varieties by farmers (every 3-4 seasons), it is difficult to keep their interest and get their cooperation for the entire selection process of segregating lines, which takes time to get results.

Lessons
- Support from local authorities and organizations in term of organization, management, additional funds and facilitation is very important.
- Cooperation with group/community on PPB/PVS gives better results than with individual farmers.

- Farmer Field School and Farmer Field days on PPB/PVS are good ways to motivate the farmers' participation at community level.
- Farmers conserve and maintain PGR diversity to meet their needs related to home consumption, market economy and adaptation to local environments and farm resources.
- Biodiversity development should be considered on both temporal and spatial bases at species, crop and agro-ecosystem levels. PPB/PVS increases PGR at a genepool level and not at specific varieties level.
- *In situ* and *ex situ* conservation and development are complementary.
- Biodiversity in the Mekong Delta is currently under pressure but integrated farming systems and PGR diversification could help to correct the situation.

Participatory approaches are very important and efficient for crop improvement at community level in Vietnam. PPB and PVS are the key tools for crop improvement. Successful results from farmer selections have strongly proven that these approaches are right. This experience is very useful for national crop improvement programmes.

References

CBDC. 1996. Annual CBDC project report of 1996.

CBDC. 1997. Annual CBDC project report of 1997.

CBDC. 1998. Annual CBDC project report of 1998.

COWI. 1999. Seed sector study, Vietnam. Volume 1: Findings, Conclusions and Recommendations. Danida/MARD draft report. March 1999. Hanoi, Vietnam

De, Nguyen Ngoc. 1997. Data collection and analysis in the Mekong Delta Community Biodiversity Development and Conservation Project of Vietnam. Pp. 29 *in* Strengthening the scientific basis of *in situ* conservation of agricultural biodiversity on-farm: Options for data collecting and analysis (D. Jarvis and T. Hodgkin, eds.). Proceedings of a workshop to develop tools and procedures for *in situ* conservation on-farm, 25-29 August 1997, Rome, Italy.

De, Nguyen Ngoc and Huynh Quang Tin. 1998. Participatory plant breeding in Vietnam CBDC project. Technical report to CBDC Regional Co-ordination Unit.

Vo-Tong Xuan. 1993. Present status of Agricultural Extension in Vietnam. Paper presented at the first Southeast Asia workshop on formulation of project proposals on technology transfer for major food crop production, FAO and UAF, HoChi Minh city, 6-9 Dec. 1993.

Sthapit, B.R., K.D. Joshi and J.R. Witcombe. 1996. Farmer participatory crop improvement. III. Participatory plant breeding, a case study for rice in Nepal. Exp. Agric. 32:479-496.

Witcombe, J.R., A. Joshi, K.D. Joshi and B.R. Sthapit. 1996. Farmer participatory cultivar improvement. I: Varietal selection and breeding methods and their impact on biodiversity. Exp. Agric. 32:445-460.

13. Mass selection: a low-cost, widely applicable method for local crop improvement in Nepal and Mexico

Bhuwon Sthapit, Pratap Shrestha, Madhu Subedi and Fernando Castillo-Gonzales

Introduction

Many farmers in developing countries plant a number of landraces and farmer-selected modern cultivars of major and minor crops for food security. Landraces herein are defined as farmers' varieties that have not been improved by a formal breeding programme. Traditionally, farmers are dependent upon their own skills and resources to develop crop varieties to suit their needs and environments. Farmers have selected present-day landraces from naturally available genetic variability. This process of local crop development is still an important institution of plant breeding in most marginal environments of developing countries. This article documents case studies from Nepal and Mexico, which used mass selection as a participatory method for crop improvement aimed at strengthening sustainable seed-supply systems and stable food production strategies of the rural farming communities. Though plant breeders often consider mass selection as an obsolete breeding method because of the low response to selection (Soleri *et al.* 1999), farmers from many generations have employed this method. Farmers' participation, particularly when it is collaborative selection of preferred crop population, can result in a much greater diversity in their fields and provide a broad range of promising genotypes (Witcombe 1999). Experiences show that with some training to improve their skills, farmers can select crop varieties that retain locally important adaptive and quality traits, and improve the productivity of crops in the process.

Local crop development and response to selection

New landraces will continue to evolve under a changing environment as long as the process of generating new diversity and selecting desired populations exists in farmers' fields. This process has been termed "evolutionary breeding" (Berg *et al.* 1991), "local crop development" (Hardon and de Boef 1993) or "*in situ* (on-farm) conservation" (Bellón 1996; Jarvis *et al.* 1998). Often, landraces are selected and maintained as populations rather than pure line cultivars. As a result they exhibit great intra-specific diversity as well as inter-specific diversity in some cases. Through traditional selection, knowingly or unknowingly farmers try to create better populations that show the most desirable performance for specific traits or local adaptation. These are morphologically distinct populations of a crop, which have a degree of genetic integrity maintained in part by natural selection and in part by human selection.

Mass selection is a common technique used by farmers to develop local crop diversity. Farmers, like plant breeders, have their own multistage selection criteria by which they evaluate new cultivars. Whether the source of seed is landraces or modern cultivars, farmers make decisions in the process of planting, managing, harvesting and processing their crops that affect the genetic make-up of the crop populations. The continued use of on-farm selection process of crop cultivars contributes to stable food production and income, especially in marginal environments where the impact of modern varieties is either limited or less effective. Experience suggests that farmers are capable of selecting varieties with good commercial qualities (Kornegay *et al.* 1996), multitraits such as yield, biotic and abiotic stress tolerance, taste, etc. (Sthapit *et al.* 1996; Witcombe *et al.* 1996) and yield performance (Sperling *et al.* 1993). Farmers, particularly women farmers, are also good at selecting for post-harvest quality traits, as they and their families are the immediate judges of qualities in new cultivars. However, they often lack knowledge and skill to balance the selection pressure that generates crop varieties with an optimal mix of productivity and quality traits. Very few research and crop improvement programmes have tried to understand local seed selection practices, identify the weakness of informal seed production systems and improve methods so that response to selection is better or aimed in the desired direction.

Cases of rice and maize from Nepal

Rice

In Begnas village of IPGRI's *in situ* project site, 173 households grow rice, maintaining more than 69 rice cultivars. More than 69% of the area is covered by landraces, and more than 97% of the rice area is planted with the farmers' own seed (Baniya *et al.* 1999). Most of the farmers (85%) change seed lots or cultivars regularly and about 49% follow this practice every year or every second year. Exchange of rice seed for either rice seed or rice grain is the predominant practice of seed flow and only a few farmers provide seed as a gift to their relatives within and outside their villages. The farmers' perception is that regular replacement of seed lots or cultivars helps increase the production. The study tried to understand whether farmers deliberately or unknowingly practise panicle selection of the preferred rice varieties they grow year after year.

Studies revealed that two special operations, i.e. rogueing and seed selection, are practised by the majority of farmers. These two operations have an important role in changing the population structure in terms of gene frequency. Most farmers decide on the best plot of standing rice for seed production, mark it and harvest the plot for seed purposes. Farmers remove all the unwanted plants from the bundle and thresh either by beating it on a hard surface or objects or by rubbing it with their legs/hands. Almost all farmers select seed before the harvest of the crop. When a particular landrace population is mixed up with off-type plants, few farmers do panicle selection to purify seed lots after every 3 to 4 years. The better-looking panicles are selected directly from the population. So, this process increases the gene frequency of desired high yield and preferred traits. Traits such as high tillering, non-lodged plants, high grain density and attractive grains are the main criteria used by farmers for selecting panicles or seed production patches from the rice fields. Lodging is the main problem in Begnas village where 69% of farmers still grow local cultivars. Though genetic gain to selection is low, farmers of this village have been found to select seeds from plots without lodged plants. Seed viability and vigour may be the primary objective of such selection. There are many cases in Nepal where farmers do keep part of the harvested grains as seed at the beginning of the season or search for leftover rice seedlings when they are short of planting materials.

Agromorphological criteria are always used at some stage in the process of seed selection. When and who makes the decisions as to which agromorphological criteria are selected may vary widely across households in a community or even within a household. The advantages of using mass selection are that the selected population still has adaptive traits of landraces and thus is more stable in marginal environments than crop varieties selected from pure line breeding techniques.

Maize: case 1

During 1970-85, Lumle Agricultural Research Centre (LARC) carried out a massive village-based maize seed production and training programme in the 25 Village Development Committees (VDCs) of the Western Hills of Nepal. The objective of the programme was to locally multiply seeds of improved maize varieties so that farming communities benefited from improved access to quality seeds at relatively lower price (because of reduced transportation cost). During the last 20 years (1973-92), 93.2 t of modern composite maize seed were distributed in 25 VDCs of Kaski, Parbat and Myagdi districts (Jaiswal *et al.* 1992). Along with the improved seeds, farmers also received on-the-spot training on mass selection of maize. The participatory approaches used in the process are:

- Distribution of minikits[4] of modern maize varieties supplied by the national breeding programmes for on-farm testing by farmers
- Village-based field staff select farmers for village-based seed multiplication programme and assist them to multiply seeds of preferred improved varieties within isolated areas of the village
- "On the spot" training is provided to farmers' groups by a combined team of field staff and station-based researchers in each village to impart knowledge and skills on seed multiplication and mass selection
- Monitoring of post-harvest seed selection through farmer groups
- Re-distribution of maize seed to farming households within and outside the village.

The results of on-farm trials showed that the modern cultivars of maize performed significantly better than local landraces in early years but, surprisingly, in subsequent years the yield difference between modern maize varieties and local landraces was not significant. This result was consistent across all 25 VDCs and, therefore, raised concern on the effectiveness of the national maize-breeding programme. The analysis of the situation later revealed that the result is largely due to the deployment of new maize diversity and regular mass selection training provided to women and men farmers. **The spontaneous crossing between improved variety and local landraces, and application of selection pressure toward desired traits in subsequent generations appears to have upgraded the productivity of local landraces as well.** An impact study, conducted by LARC with a sample size of 1800 households in these VDCs, has indicated that 53% of farmers have adopted modern cultivars and that the yield level of maize, including that of local varieties, has also increased (LARC 1996).

Maize: case 2

The maize production practices of farmers of Gulmi and Argakhanchi districts in Western hills of Nepal provide another example of where farmers have traditionally been using mass selection techniques to maintain the productivity of their local maize varieties. A recently conducted study (Subedi and Shrestha 1999) in parts of Gulmi and Arghakhanchi districts has revealed that, through years of mass selection, farmers have developed a popular variety, which is locally called *Thulo Pinyalo* (literally meaning "big yellow"). This is a widely grown variety and covers around 85% of the total maize area of the region. Though Thulo Pinyalo is a tall-statured variety (plant height was recorded as high as 6.5 m), it is most preferred by the farmers in the area because of its high yield potential, good culinary traits, higher grit recovery, insect (storage) and disease tolerance, fodder quantity and palatability, and its adaptation to the local management practices.

The secret of all the good characters in *Thulo Pinyalo* is its broad genetic base contributed by many improved genepools of diverse origin. This variety has resulted from farmers' continuous selection work in a heterogeneous population (Fig. 13.1). The heterogeneity in local population has been brought about by the introduction of exotic varieties by various programmes and/or organizations in the area. For example, an American variety (real name not known), donated and distributed as food grain under USAID's 'Food for Work' programme, was introduced in the area about 36 years ago. Farmers saved seeds and planted in the subsequent years. Similarly other improved (both released and pipeline) varieties, developed through formal breeding programmes in Nepal, have been introduced in the area in the form of minikits and under extension programmes of the government during the last 20 years. Thus the exotic genepools have continuously contributed their genetic background to farmers' varieties.

---

[4] A small seed kit of modern cultivars distributed free by the formal research system to assess the popularity of cultivars under farmers' management conditions. The term was first coined by the late Bill Golden to introduce new varieties of rice in Philippines in the late 1960s.

Fig. 13.1. Nepalese farmers selecting local maize seed for planting.

The spontaneous crossing between exotic and local varieties grown in the area resulted in a heterogeneous population in the subsequent generation. This heterogeneous population, in turn, has offered farmers an opportunity to make selections for better ears and consequently develop a new population with farmers' preferred traits. In this way, farmers of Gulmi and Arghakhanchi have bred *Thulo Piyalo* maize, which satisfies most of their requirements for a variety. However, this variety has one poor character. It is highly prone to lodging and causes up to 80% losses in bad years. Its excessive plant height appears to be one of the major contributing factors to this problem. The study in the area has revealed that farmers select harvested cobs and not the standing plants for seeds for next season (Subedi and Shrestha 1999). They keep seeds from big and long cobs with bold grains. Since tall plants usually have big and long cobs, the seed selection pressure is unknowingly toward taller plants. Farmers were found to be unaware of this. With support from SWP's PRGA programme, LI-BIRD (a local NGO) is currently working with farmers' groups in two VDCs of Gulmi district to address this problem. Under this programme, farmers are provided with improved maize germplasm and training in mass selection to enhance their capacity to breed new *Thulo Pinyalo* with optimal plant height.

Case of maize from Mexico

Traditional agriculture in Mexico depends on native genetic diversity for crops such as maize, beans, squash and chilli pepper. It means that most of the diversity for these crops is still in the farmers' fields; improved seeds are planted only in about 15% of the maize-cultivated area, mostly in the western and northwestern states of Mexico. To improve productivity and other traits in the landraces of maize, farmers' participatory approaches using mass selection method

have been employed in recent years. Farmers have been found to improve productivity of their maize varieties and enhance other traits by making some modifications to the way they do seed selection. Starting with those farmers who own the best-yielding populations, mass selection can help to obtain further productivity gains. Following are three broad steps used for such work in Mexico:
- Identification of landrace forms and documentation of their uses
- Detection of best-yielding maize landrace populations on-farm
- Mass selection in farmers' fields.

An on-farm plot of 50 m² is selected from which a farmer is able to select plants with preferred agromorphological characteristics such as shape, size and plant health. From selected plants, large ears and seeds are selected for the next season. Selection done through each subplot may offer expected gain in the order of 2% per generation (Castillo *et al.* 1999). As raw landraces are very variable, gains in the early generations may be greater. Preliminary assessment for three cycles of selection done over five different populations in the Mexican highland yielded 20% gains. If actual gains were half of this figure, doing mass selection among the local maize population is still a worthy effort to pursue. Furthermore, any farmer, with very little technical assistance for a few generations, can carry out this kind of participatory maize improvement without much additional cost.

Conclusion

Through traditions, farmers have been using mass selection as one of the most common methods to improve productivity of and combine traits of their preferences in the crops they grow. Farmers possess considerable knowledge, skills and experiences that enable them to select the right plants and/or seed for subsequent years. This makes mass selection an effective and cheap means of crop improvement in the farming communities of rural and remote areas. This simply requires providing access to new germplasms, with desired traits, from which farmers can select and maintain a population of their choice. Additional breeding knowledge and skills can be transferred to the farming community through village workshops, where farmers and scientists can share their knowledge and skills. In remote areas and where seed sources are mainly from seed retention, mass selection plays a significant role in participatory plant breeding. This further empowers farmers and farming communities to develop and retain crop varieties that meet their local requirements.

References

Baniya, B.K., A. Subedi, R.B. Rana, B.R. Sthapit, D.K. Rijal, C.L. Poudel and S.P. Khatiwada. 1999. Informal Rice Seed Supply and Storage Systems in Mid-hill of Nepal. Paper prepared for "The Third Global Participants Meeting on Strengthening the Scientific Basis of *In-situ* Conservation of Agricultural Biodiversity on-farm: 5-12 July 1999. Pokhara, Nepal.

Bellón, M.B. 1996. On-farm conservation as a process: An analysis of its components. Pp. 9-22 *in* Using Diversity: Enhancing and Maintaining Genetic Resources On-Farm (L. Sperling and M. Loevinsohn, eds). Proceedings of a workshop held on 19-21 June 1995, New Delhi, India. IDRC: New Delhi.

Berg, T., A. Bjornstad, C. Fowler and T. Skroppa. 1991. Technology options and the gene struggle. A report to the Norwegian Research Council for Science and Humanities. Development and Environment No. 8, NORAGRIC Occasional Paper Series C. Agricultural University of Norway.

Castillo, F., L.M. Arias, R. Ortega and F. Marquez. 1999. Adding benefits through PPB, seed networks and grassroots strengthening. Proceedings of "The Third Global Participants Workshop" to develop tools and procedures for *in situ* conservation on-farm, 5-12 July 1999, Pokhara, Nepal.

LARC. 1996. The adoption and diffusion and incremental benefits of fifteen technologies for crops, horticulture, livestock and forestry in the Western Hills of Nepal. LARC Occasional Paper 97/1. Lumle Agricultural Research Centre, Pokhara, Nepal. pp. 30.

Hardon, J. and W.S. de Boef. 1993. Linking farmers and breeders in local crop development. Pp. 64-71 *in* Cultivating Knowledge Genetic Diversity, Farmer Experimentation and Crop Research (W.S. De Boef, K. Amanor, K. Wellard and A. Bebbington, eds.), ITP, London.

Jaiswal, J.P., A. Subedi and K.J. Gurung. 1992. Seed production on cereals, grain legumes, oilseed and potato crops in LARC's command area: Existing system and review of the past work. Review Paper No. 1992/5. Lumle Regional Agricultural Research Centre, Pokhara, Nepal.

Jarvis, D., T. Hodgkin, P. Eyzaguirre, G. Ayad, B.R. Sthapit and L. Guarino. 1998. Farmer selection, natural selection and crop genetic diversity: the need for a basic dataset. Pp.1-8 *in* Strengthening the scientific basis of *in situ* conservation of agricultural biodiversity on-farm. Options for data collecting and analysis (D. Jarvis and T. Hodgkin, eds.). Proceedings of a workshop to develop tools and procedures for *in situ* conservation on-farm, 25-29 August 1997, Rome, Italy.

Kornegay, J., J.A. Beltran and J. Ashby. 1996. Farmer selections within segregating populations of common bean in Colombia. Pp. 151-160 *in* Participatory Plant Breeding (P. Eyzaguirre and M. Iwanaga, eds.). Proceedings of a workshop on participatory plant breeding, 26-29 July 1995, Wageningen, The Netherlands. IDRC /FAO /CGN/ IPGRI.

Sperling, L., M. E. Loevinsohn and B. Ntabomvra. 1993. Rethinking the farmer's role in plant breeding: Local bean experts and on-station selection in Rwanda. Exp. Agric. 29:509-519.

Sthapit, B.R., K.D. Joshi and J.R. Witcombe. 1996. Farmer Participatory Crop Improvement: III Participatory Plant Breeding: A Case Study for Rice in Nepal. Exp. Agric. 32:479-496.

Soleri, D., S. Smith and D. Cleveland. 1999. Evaluating the potential for farmer-breeder collaboration: A case study of farmer maize selection from Oaxaca, Mexico. AgREN Network Paper No. 96a. ODI Agricultural Research and Extension Network, July 1999.

Subedi, M. and P.K. Shrestha. 1999. Site selection report of Farmer-led Participatory Maize Breeding Programme for the Middle Hills of Nepal, 1998. LI-BIRD, Pokhara, Nepal.

Witcombe, J.R., A. Joshi, K.D. Joshi and B.R. Sthapit. 1996. Farmer Participatory Cultivar Improvement. I: Varietal Selection and Breeding Methods and their Impact on Biodiversity. Exp. Agric. 32:445-460.

Witcombe, J.R. 1999. Participatory Plant Breeding and Broadening the Genetic Base of Crops. *In* Broadening the Genetic Base of Crops (D. Cooper, T. Hodgkin and C. Spillane, eds.). CAB International, UK. (in press)

14. Understanding farmers' knowledge systems and decision-making: participatory techniques for rapid biodiversity assessment and intensive data plot in Nepal

Ram Rana, Pratap Shrestha, Dipak Rijal, Anil Subedi and Bhuwon Sthapit

Introduction

Farmers' decision-making on management of crop diversity on-farm is a complex phenomenon influenced by internal as well as external factors of the farming household. A clear understanding of the rationale of the farmers' decision-making process is essential for devising sound conservation strategies. In this regard, participatory approaches such as rapid biodiversity assessment (RBA) and intensive data plots (IDP) play a key role in understanding the varietal dynamics at household level and the management of biodiversity on-farm in different agro-ecological and socioeconomic settings. The RBA and IDP, although they appear contrasting, actually complement each other in terms of generating information from a general to a more specific level. RBA helps to understand a broad species-level diversity at household and community levels, whereas IDP tries to understand and document the knowledge systems of farmers' decision-making about crop diversity.

The RBA method of assessing biodiversity has been used in an IDRC-funded family nutrition related biodiversity project in Nepal. A number of RBA techniques have been used to assess biodiversity existing at community and household levels at the project sites. The information documented has been used to design agrobiodiversity enhancement interventions as well as to set indicators for monitoring and evaluation of changes in the existing agrobiodiversity. Similarly, the IDP method has been used in a global project entitled "Strengthening the scientific basis of *in situ* conservation of agriculture biodiversity on-farm", coordinated by the International Plant Genetic Resources Institute (IPGRI), Rome, Italy. The IDP technique is employed to generate plot-level information on the varietal dynamics, management practices for different varieties/landraces across socioeconomic factors, economics of producing different landraces and the implication of these factors for the genetic diversity of landraces.

Rapid biodiversity assessment

Rapid biodiversity assessment (RBA), among many others (listed in Guarino and Friis-Hansen 1995), is a qualitative analytical tool that uses participatory and rapid information-collecting techniques in assessing the biodiversity of a location. RBA uses a number of participatory appraisal techniques to suit different levels of biodiversity assessment. Depending on the objectives and scope of biodiversity assessment, RBA can be done at both community and household levels. Similarly, based on the nature of the work, RBA can include a wide range of elements/components of biodiversity. For example, the components of RBA on agrobiodiversity could be species and varieties/breeds of food crops (cereals, legumes, potato and oilseeds), vegetables, fruits, livestock (including birds), fodder and forage grasses, spices and condiments, medicinal and aromatic plants, ornamental plants, and others (coffee, *lapsi*, etc.).

Community-level rapid biodiversity assessment

Community-level RBA involves documentation of biodiversity at a general level (i.e. community as a whole), without much consideration given to inter-household differences in the use and maintenance of the existing biodiversity. In this case, the main focus lies on the total biodiversity in a community as a reflection of its location (access, altitude, aspect and other agro-ecological conditions), sociocultural tradition and development interventions. The following rapid assessment tools have been used for community-level RBA.

Focus group discussion

The Focus Group Discussion (FGD) involved the following steps.

1. Selection of key informants and organizing a meeting. In consultation with the field staff and village leader, about 10 to 15 experienced (involved in farming for more than 5 years) male and female farmers are selected and invited to a meeting.

2. Clarifying objectives and building rapport. Before initiating actual discussion, participants (both farmers and facilitators) introduce themselves. The invited farmers are then briefed on the purpose of the meeting and importance of information on the state of biodiversity in the community for the project. This also helps to build rapport with the key informants.

3. Participatory discussion. In an informal atmosphere, key informants are urged and encouraged to share their experiences and provide information on the state of agrobiodiversity in the community. A tabular checklist (semi-structured questionnaire) is used to document the extent and trends of agrobiodiversity (both plants and animals) in the community (sample presented in Table 14.1).

As is obvious from Table 14.1, the focus group discussion documents inter- and intra-species diversity, their extent of diversity (% households growing and % area under them), their distribution among different social classes (ethnic and wealth categories), their increasing or decreasing trends in the community and the associated reasons for such trends.

4. Biodiversity time line. The biodiversity time line is another rapid participatory tool, which helps tracking changes in the biodiversity of a community over time. After FGD, the same group of key informants is used to draw a time line to record historical changes in agrobiodiversity in the community using a data format presented in Table 14.2.

The biodiversity time line thus reveals the dynamics of biodiversity in a community. It clearly shows frequency of introduction and/or replacement or disappearance of plant and animal species in a community, and documents reasons or causes for such changes.

Household-level rapid biodiversity assessment

Household-level RBA entails more detailed and specific documentation of biodiversity at individual household level taking into account a number of factors like ethnicity, resource endowments and access to external opportunities (such as market, off-farm employment, etc.). Here a biodiversity mapping tool, used to assess agrobiodiversity at the household level, is presented.

Biodiversity mapping

The biodiversity mapping employs mapping techniques to produce maps that show biodiversity at household level. These maps show spatial distribution of plant species and farm resources around the farm and their association with each other, and help to produce a list of plant and animal species as well as their varieties and breeds. Biodiversity mapping and recording of associated information involved the following steps.

1. Selection of representative households. To avoid the cost of mapping biodiversity of all households of the community, representative households are selected for the purpose. The representative households can be selected based on the factors of interest to the study, for example ethnicity,

Table 14.1. Agrobiodiversity situation at the study village (a sample tabular checklist)[†]

Crops/varieties	% HHs involved	% area covered	Reporting ethnic groups	Reporting wealth class	Diversity ↑↓ trends	Reasons for ↑↓ trends
1. Crop-1 (e.g. Maize)						
Variety-1 (name)						
Variety-2 (name)						
2. Crop-2 (name)						
Variety-1 (name)						
Variety-n (name)						
n. Crop-n (names)						
Variety-n (name)						

[†] HHs represent households, and ↑↓ indicate increasing (↑) and decreasing (↓) trends.

Table 14.2. Biodiversity time line recorded at Arba village, ward 4 of Arba Village Development Committee (a hypothetical example)

Year of events	Events relevant to biodiversity	Reasons/causes for events
1980	Area under summer maize and blackgram drastically decreased	*Bari* lands (upland) were converted to *khet* land (rice-growing irrigated lands)
1990	Introduction of Mansuli variety of rice and its gradual expansion	Government extension agency promoted the variety with free seeds
1995	Disappearance of aromatic variety of sponge gourd from the village	Young fruits were eaten by monkey and seeds could not be collected

wealth categories, education status of household heads, etc. Alternatively once the number of combination of factors is decided, selection of households also can be done on the basis of proportionate random sampling.

2. *Determining the elements of biodiversity mapping.* The number of elements in a biodiversity map depends on the objectives of the study. A biodiversity map should generally consist of:
- area, shape and orientation of the homestead and/or farm land
- location of house, animal shed, bird housing, compost pit, water taps, toilets
- location and distribution (relative placement) of vegetables (both annual and perennials), fruits, fodder trees and grasses, and medicinal and aromatic plants in the homestead and/or in the whole farm
- distance of field crop lands from the house
- a list of field crops and cropping patterns for each parcel of field crop land.

3. *Deciding on the level of mapping.* Depending on the scope of the study, biodiversity mapping can be done at (a) homestead and/or (b) whole farm level. Homestead biodiversity mapping records biodiversity around the farmhouse and largely includes vegetables, fruits, fodder trees, and other resources like animal shed, compost pit, water tap, etc. A sample spatial

homestead biodiversity map is presented in Figure 14.1. On the other hand, whole farm mapping also includes distantly located field crops, fodder trees and grasses, and orchard. In addition to a spatial whole farm biodiversity map (similar to Fig. 14.1), a biodiversity transect map also can be drawn to capture/document a detailed spatial diversity of a farm (sample map in Fig. 14.2).

4. *Drawing maps.* Biodiversity maps can be drawn either by researchers themselves in consultation with the household head and the family members or by the household members themselves with facilitation from the researchers. The following procedures can be adopted in drawing biodiversity maps:
- Explain purpose of the mapping to the household head and/or family members and their role in drawing the maps.
- Walk around the house and the farm making a transect map and observe biodiversity, discuss with the family members and take small notes for later reference in drawing maps.
- Draw maps of homestead and/or whole farm, similar to resource map, including elements mentioned above and showing their relative location.
- A number of maps can be drawn to cover larger area of the farm or to show more detailed configuration of biodiversity or to capture different seasons, and later superimposed to produce a single more detailed biodiversity map.

5. *Documenting associated details.* After drawing maps, a detailed listing of species and their varieties/breeds, and of the family's knowledge associated with the biodiversity is prepared. A data format presented in Table 14.3 is used for the purpose. This is important as it is not possible to show every minute detail, especially names of each plant species grown in different seasons, on the biodiversity maps.

6. *Synthesis for visible patterns.* A careful analysis, through contrast comparisons and analysis of difference, of biodiversity maps reveals clear patterns in terms of similarity and differences in agrobiodiversity across mapped households. These similarities and differences then can be correlated with ethnicity, wealth categories, education levels of household heads and with other related factors. Figure 14.1 clearly shows differences in biodiversity resources due to differences in wealth category and ethnicity, and points out possible intervention areas.

Observations in using RBA
The following observations have been made in the use of RBA tools in this case.
- RBA can be used as a rapid and easy method for getting local participation and building rapport.
- Community and household level RBA can be combined/integrated to give a more contextual picture of the biodiversity situation of a location.
- Area, shape and orientation of homestead, and distribution of plant species and other resources are important in biodiversity mapping as they indicate the current intensity of biodiversity and possible intervention areas.
- Participation of women and elderly people is crucial in getting reliable and precise information as this group of people has considerable experience in the management of local PGR.
- The outcomes of RBA are not only useful for planning PGR management strategies but are equally important for establishing a baseline to monitor changes over time.

Field implementation and testing of these RBA tools and many others in different conditions will help to better understand the situation of local PGR and, as a result, to design appropriate PGR management strategies.

Local PGR management and participatory crop improvement

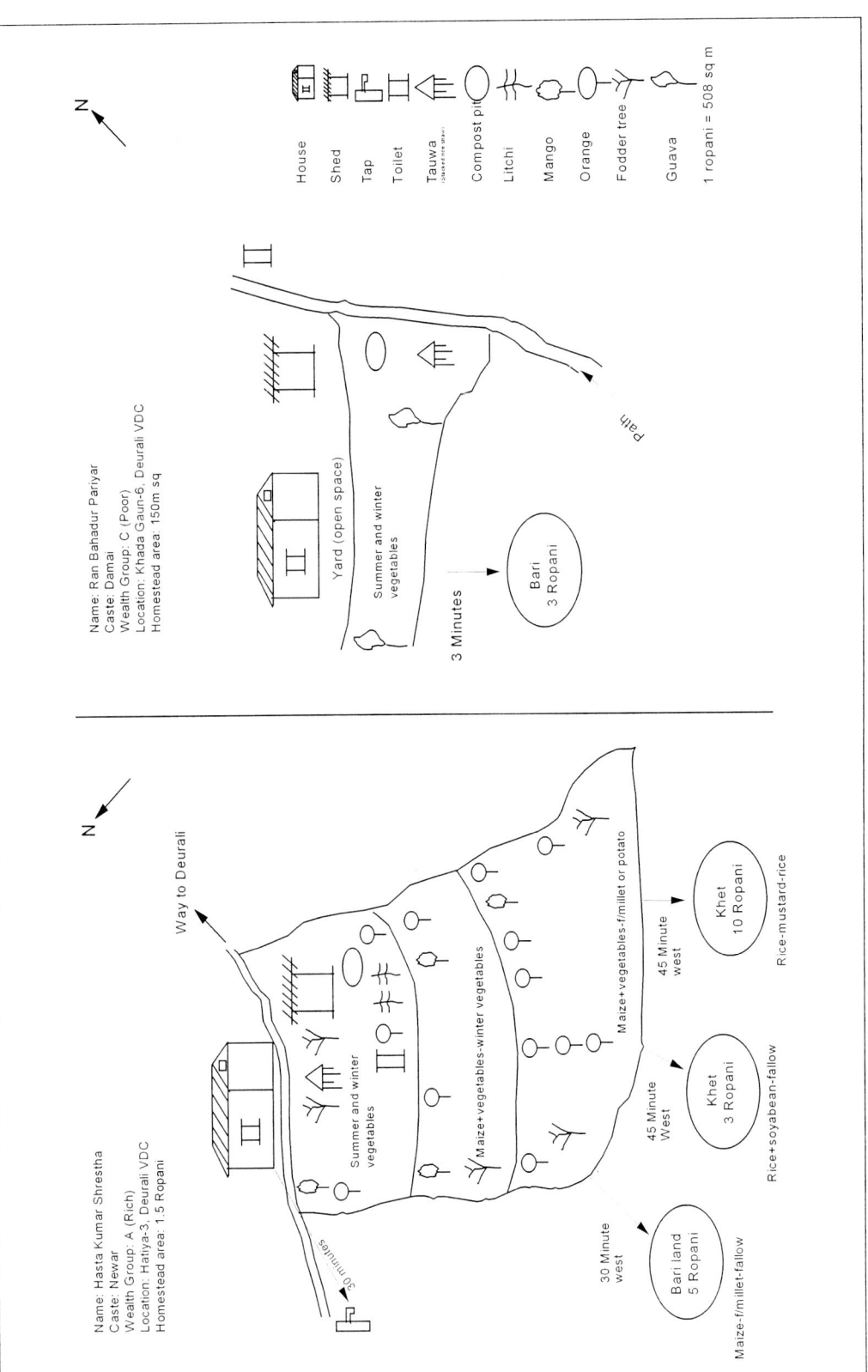

Fig. 14.1. Homestead biodiversity maps.

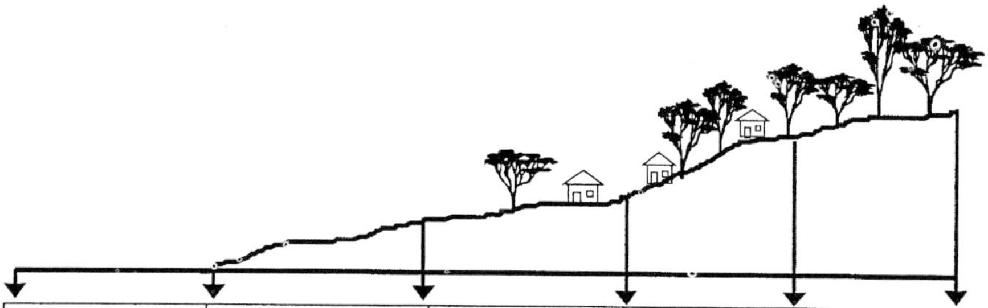

Land types (Production unit)	Khet land (Irrigated lowland)	Bari land (rain-fed upland)	Homestead (Home garden)	Kharbari (Pasture land)
Land area (ha)	3	5	1.25	2
Distance to farm house (minutes)	30	15	0	5
Soil types	Clayey	Loam/sandy loam	Red soil	Stony silty
Water regimes	Irrigated	Rain-fed	Rain-fed	Rain-fed
Crop diversity (field crops)	Rice, wheat, maize, lentil	Maize, finger-millet, soyabean, blackgram, rice-bean, potato, mustard, barley	Maize, finger-millet, potato	None
Vegetable and spices diversity	None	Beans, pumpkin, taro, yam	Beans, cucumber, pumpkin, taro, yam, radish, rayo, gourds, ginger, turmeric	Sponge and bottle gourd, pumpkin, taro, yam
Fruits/trees diversity	None	Orange, guava	Orange, guava, pineapple, peach, plum	Pineapple, peach
Herbs and aromatic plant	None	None	Basil, *Marathi*	None
Diversity problem	Lack of winter irrigation	Irrigation, new crop species	None specific	Low diversity, space under-utilized
Opportunities for diversity	Drought-tolerant crop varieties	Fruit and fodder tree diversity	Perennial vegetables and fruits	Grasses and fruits

Source: PRA handouts prepared by B. Sthapit for PRA training in Nepal and Vietnam)

Fig. 14.2. A whole farm biodiversity transect map (sample).

Table 14.3. List of species shown in the biodiversity map and associated knowledge base

Crop types/ crops	Variety/ breed name	Variety/ breed type (local/improved)	No. of varieties/ breeds	Parts used as food	Special food value	Seed sources
1. Vegetables:						
pumpkin						
taro						
2. Fruits:						
3. Other						

Intensive Data Plots

Intensive Data Plots (IDP) are mainly employed to understand the farm economics and farmer-managed practices at household level on a regular basis. The technique, as reported by Hobbs *et al.* (1996), was a very effective way of participatory recording and analysis of on-farm activities in a rice-wheat system, which provides an insight into farmers' behaviour and decision-making processes. In the *in situ* project the purpose of this method is to understand knowledge systems of farmers' decision-making about crop diversity by intensive monitoring of a plot where a farmer grows a specific variety. All natural, socioeconomic and human-managed factors of the plots and household are measured and related to the existing crop biology (Fig. 14.3). Specific objectives of IDP include:

- keep an inventory of landraces/varietal diversity on a specific plot at household level
- document indigenous knowledge systems associated with landraces
- calculate the cost of production and return (Cost Benefit Analysis) of landraces/varieties
- record farmers' management practices on different landraces/varieties
- diagnose farmers' problems and constraints, and explore opportunities for improvements.

The processes/steps involved in IDP are easily made participatory without compromising on the quality of data required for scientific rigour that is demanded by the project. Application of the IDP technique within the *in situ* project is rather recent. The technique is still evolving, and the following sections concentrate on the processes adopted by the Nepalese *in situ* project team.

1. Brainstorming session by National Multidisciplinary Group (NMDG). A group discussion by NMDG members on methodological aspects facilitates refining the methodologies for the IDP exercise. At the same time, thematic leaders agree on the types of data to be collected at plot level, and involvement of different members in the process is finalized.

2. Transect walk by Multidisciplinary Team. The multidisciplinary team comprising Thematic Leaders (TLs), Site Coordinators (SCs), Field Staff (FSs) and representative farmers makes a transect walk of the village to identify major domains (production environments) based on land use, soil type, water regimes, etc. and to locate specific landraces or varieties for different domains. Once landraces are identified for each domain, farmers and researchers jointly select a few landraces for the IDP exercise. Households growing the selected landraces are identified and their approval is sought for participation in the programme.

3. Selection of farmers. While identifying and selecting participant farmers, the expert knowledge of field-based staff and key informants are utilized. The selection of participants is based on the following criteria:

- Participating farmers from different wealth categories are selected. In order to capture the variability in socioeconomic factors (farm management practices, input levels, credit and market access, etc.) across wealth categories, participants are selected from all wealth groups (based on their own criteria).
- Literate farmers are selected. The participant farmers do all the recording themselves, so literate farmers (someone from the family) are selected.
- Selection of cooperative farmers. Since participant farmers are engaged in regular data keeping and recording, their commitment and cooperation are essential.
- Cultivator farmers with some years of farming experience. Farmers with 'true farming experience' who have been directly involved in cultivation of land for some years participate in the IDP exercise since they can contribute by providing relevant information.

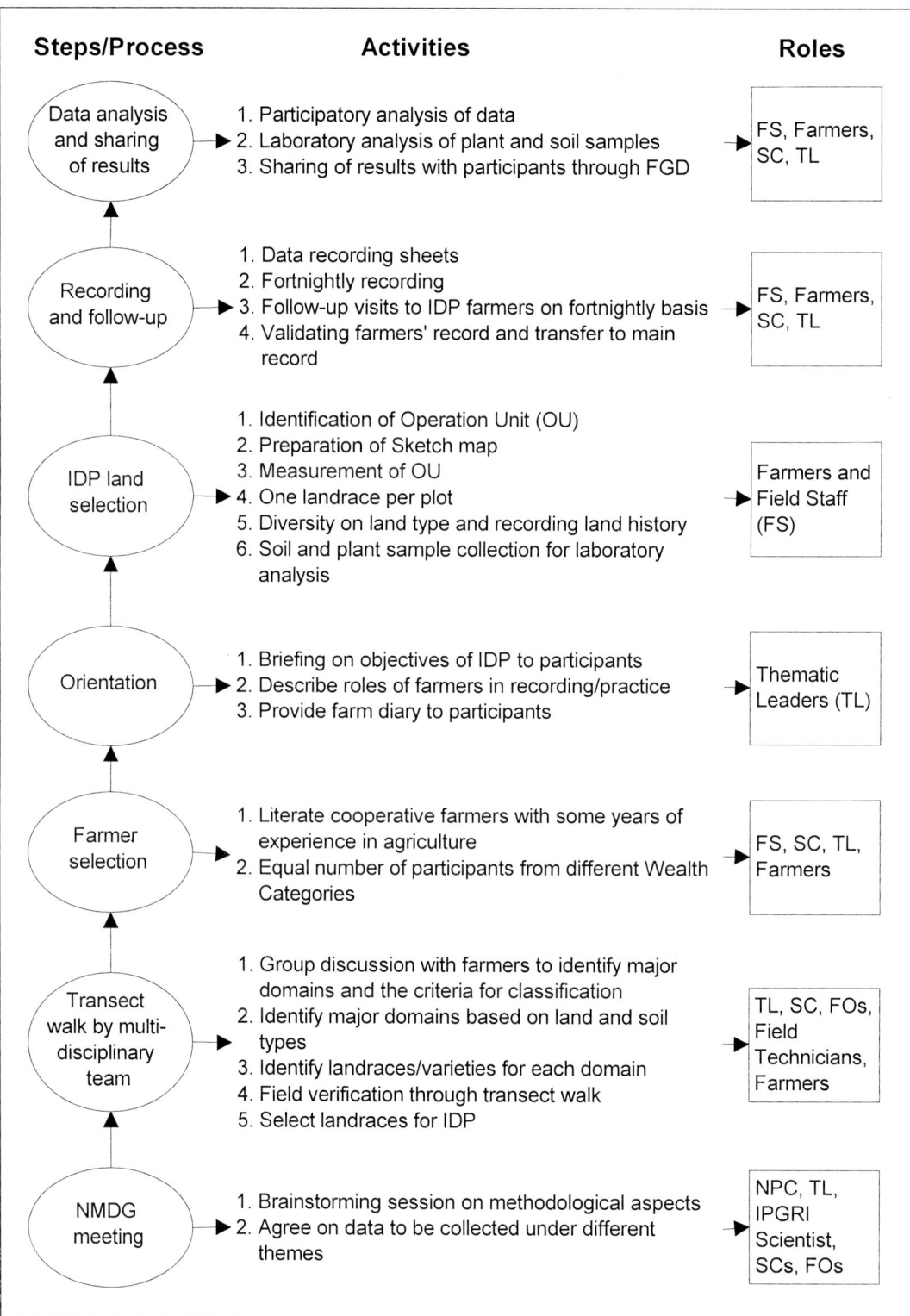

Fig. 14.3. Intensive data plots.

4. *Preliminaries/Orientation.* Although some level of orientation is provided to potential participants for facilitating decision-making in the exercise, the detailed orientation is carried out only with the selected participants and comprises the following:
- Briefing on objectives of IDP to participant farmers. A formal workshop is organized to provide introduction to the subject matter as well as to highlight the benefits of participating in the programme and to explain their roles and responsibilities in the whole process.
- Role of farmer/recording system. At this stage participant farmers are made aware of what their roles and responsibilities are regarding the record-keeping of all the activities pertaining to IDP. Detailed discussions are held, and a dummy exercise is run to make participants confident about the exercise.
- Provide farm diary. Once participant farmers feel comfortable with the record-keeping system, they are provided with a diary and pen to record all the activities in relation to IDP.

5. *IDP land selection.* Actual implementation of the IDP exercise begins by joint selection (participant farmer and field staff) of an operational unit of land, measurement of the land area and recording of the land history.
- Isolated/identifiable operational unit. It is important to have a common understanding between participant farmer and field staff on which portion of land both are referring to. Therefore, both of them jointly identify and agree on the land unit selected for the IDP exercise, which is termed the 'Operational Unit'. The operational unit of land is something which a farmer treats as one unit, i.e. farmer would plough, sow and harvest it as one unit.
- Preparation of sketch map to identify location of IDP land. Participant farmers, facilitated by field staff, prepare a sketch map showing the exact location of the IDP within the farm.
- Measurement of Operation Unit. Size of the land is measured at the beginning of each season. This is required for computing cost-benefit analysis of landraces/varieties.
- One farmer-named variety per plot/land. Since it is not practical to do IDP on all the landraces farmers grow, the selection of landraces is the farmers' choice. The general practice is one landrace per plot.
- Diversity on type of land. The diversity in the 'Operational Unit' needs to be captured because the genetic composition of landraces might be changed as a result of local adaptation and selection over time.
- Recording of land history. Land history includes information such as cropping patterns, fallow periods, cultivation history, compost and chemical fertilizer use, application of micronutrients and other additives, irrigation, etc. In addition, direct observation of biotic and abiotic limiting factors can also be included in land history.
- Soil sample collection for laboratory analysis. Soil samples are collected from the IDP plots for nutrient analysis in the laboratory. This helps to interpret the results and validate farmers' management practices and perceptions on soil fertility.
- Plant sample collection for isozyme analysis. Crop biologists, with the help of participant farmers and field staff, select at least 30 plant samples per plot for isozyme analysis. Laboratory analyses of samples help determine the genetic diversity present within and between landraces collected from within and between plots, which could be explained by differences in agro-ecological and socioeconomic factors of the farming households.

6. *Recording and follow-up.*
- Data recording sheet. General framework with broad headings for data collecting is provided to IDP farmers for recording information.
- Fortnightly recording. Recording of IDP activities by the staff is done on a fortnightly basis. However, participants are advised to assign a fixed day in a week for writing down all the activities performed within that week.

- Follow-up visits to IDP farmers. As per the requirement, field staff have to make a weekly or fortnightly visit to the farm.
- Validation of farmers' record and transfer to main register. Before transferring the information from farm diary to main register, field staff probe on issues that are either not clear or seem unrealistic. Farmers' explanations are noted as comments on the register.

7. *Data analysis, interpretation and sharing of results.* Participatory ways of data analysis and sharing of outputs are important for sustained cooperation from the participant farmers in the programme.
- Participatory analysis of information. Data analysis and interpretation are done in a participatory manner. Individual farmers (family members) and the respective field staff analyze the information at the farm. Cost of production and returns are calculated and presented to the family. The data from different ecosites are analyzed using computer software to observe similarity and contrasting factors across sites. Outputs are presented at the field office for further sharing with field staff and respective participants.
- Laboratory analysis of plant samples. Collected plant samples are analyzed in the laboratory using different tools and the results presented to the concerned staff for further sharing.
- Sharing of information with participant farmers. Formal sharing of IDP results with the participant farmers takes place through Focus Group Discussion (FGD) and individual contact.

Conclusion

Participatory methods like RBA and IDP have been useful for this purpose. RBA is rapid and cost-effective in generating information on biodiversity and associated knowledge systems embedded in a community. In addition to other uses of this information, it can be used to narrow down information needs for IDP and to plan IDP activities accordingly. The IDP-generated data at plot level could be integrated to observe the association between landrace diversity and conditions of the plots. The advantage of IDP is that various factors which influence farmers' decision-making could be related to measured and perceived data of local crop diversity at household level with great accuracy. The RBA and the IDP thus have a huge potential and use in understanding knowledge systems of farmers' decision-making about crop diversity. They are important in formulating strategies for the conservation and use of plant genetic resources. These methods, however, can be modified to suit the local conditions and social circumstances of different locations.

References

Guarino, L. and E. Friis-Hansen. 1995. Collecting plant genetic resources and documenting associated indigenous knowledge in the field: a participatory approach. Pp. 345-365 *in* Collecting Plant Diversity: Technical Guidelines (L. Guarino, V. R. Rao, and R. Reid, eds.). IPGRI, FAO, UNEP, IUCN and CAB International, UK.

Hobbs, P.R., L.W. Harrington, C. Adhikari, G.S. Giri, S.R. Upadhyay and B. Adhikari. 1996. Wheat and Rice in the Nepal Terai: Farm Resources and Production Practices in Rupandehi District, NARC and CIMMYT.

15. Participatory approaches to a study of plant genetic resources management in Tanzania

Esbern Friis-Hansen

Introduction to study

During the 1990s, the debate about sustainable conservation and use of plant genetic resources flourished within international and national agricultural research institutions. A consensus was reached with the adoption of a Global Plan of Action for the sustainable use of plant genetic resources in 1996. Theoretically, this consensus can be characterized as a drastic departure from previous conceptual approaches, and the basis of the recommendations of the Global Plan of Action constitutes a change in paradigm. While a consensus is growing over what to do, the discrepancy between this consensus and what is done by mainstream research and development institutions remains large. In particular, studies at the **community and household levels** of the elements of plant genetic resources management in the changing socioeconomic context in developing countries have been rare.

A research programme titled 'Sustainable Agriculture in Semi-Arid Africa' was carried out in Tanzania between 1994 and 1999, as a collaboration between Danish and Tanzanian research institutes. One study carried out by the author aimed to enhance the understanding of farmers' management of local plant genetic resources in Tanzania. The rationale behind this study was to fill a knowledge gap created by the general lack of farmers' participation in the agricultural research process and lack of accountability by state institutions to farmers. Tanzanian farmers have rarely been consulted and involved in the work carried out by state-financed plant genetic resources management institutions. Plant genetic resources collecting missions have regarded farmers as potential sources of crop varieties, not as sources of knowledge regarding plant genetic resources management or partners in their conservation. Commodity plant breeding programmes have primarily been based on on-station trials and have only to a limited extent carried out on-farm trials.

The Tanzanian ministry of agriculture is currently starting to incorporate farmers into its plant genetic resources system by creating 100 village-based seed production units to complement the private seed industry, whose activity with regard to food crops is largely limited to selling hybrid maize seed from urban-based shops. The Seed Unit of the Ministry of Agriculture is contemplating using participatory approaches to determine which varieties to multiply and is prepared to include landraces to the extent that they are demanded by farmers (Mr Z. Lumbadia, Head of Seed Component, Ministry of Agriculture, Tanzania, pers. comm.). It is hoped that the participatory framework and methodology for studying farmers' plant genetic resources management, developed by this study, may be useful for future activities in Tanzania and elsewhere.

Introduction to study area and methodology

The study area is the Rhuaha river basin, which makes up the semi-arid parts of Iringa and Mbeya regions. The three villages studied were Mkulula village, situated in Ismani Division with an annual rainfall of 431 mm, Ikuwala village, situated in Iringa District with an annual rainfall of 600 mm, and Nyeregete village, situated in Mbarali District (Usangu plains) with access to traditional flood irrigation. The three villages studied represent three different types of farming systems in semi-arid parts of Tanzania and cover a range of food crops including rice, maize, sorghum, sunflower, beans, cowpeas and sweet potatoes.

The study applied a range of participatory methodologies which involved all major groups of farmers present within the three communities. Participatory research methodologies are qualitative in nature and the major emphasis in the fieldwork was on qualitative interviews and dialogues with different groups of farmers and with women and men within individual households. However, quantitative research methods, such as a random-sample household questionnaire, can fruitfully be combined with qualitative information. Such a combination enables an analysis of the relative importance of different qualitative issues and explanatory factors. In the following the different groups of methodologies applied in the study will be discussed in the sequence in which they took place, starting with purely qualitative methodologies and moving on to quantitative household surveys.

Qualitative research methodologies

- *Open and focused group interviews* were carried out during the initial stage of the study in all three villages. The open-ended group interviews were designed to enable farmers to influence the research agenda and contained a general discussion about farmers' management of plant genetic resources and about management differences between different social groups in the village.
- *Crop-specific focused group interviews/discussions* were carried out for maize, beans/cowpea, sweet potato, sorghum, sunflower, finger millet and rice. All interested farmers were invited to these group sessions, which ensured a lively discussion, as all participants were interested in the subject and many had something to contribute. Farmers were encouraged to discuss it with each other and efforts were made to register differences in opinions over issues such as use of varieties and methods of seed selection, treatment, storage and cultivation. A semi-structured questionnaire was used to generate farmers' characterization of local landraces. An example for sorghum in Mkulula vilage is shown in Table 15.1.
- *In-depth interviews with key informants* were carried out to follow up on interested issues which had surfaced during the group interviews or from other discussions with farmers. Group interviews in Mamongolo village had, for example, revealed that ethnicity played an important role in management of sorghum and maize varieties, and in-depth

Table 15.1. Characteristics of sorghum varieties in Mkulula village[†]

Variety	PN3	Msabe	Kasao	Sanyagi	Kilezilezi	Tegemeo	Mihenduno
Grain yield	good	average	good	good	good	good	average
Grain size	average	average	good	average	average	large	good
Head size	good	average	loose, large	good	large	loose, large	average
Drought tolerance	good	good	v. good	good	v. good	v. good	good
Time to maturity	v. early	medium	medium	medium	early	early	medium
Stover yield	poor	good	good	good	average	poor	average
Use of stem	poor	good	poor	good	good	average	good
Bird resistance	poor	good	good	good	good	poor	good
Disease resistance	poor	poor	good	poor	good	good	poor
Insect resistance	poor	poor	good	poor	good	good	poor
Threshing ease	good	average	good	average	average	good	average
Dehulling ease	good	poor	good	poor	average	good	poor
Grain colour	white	dark red	white, black	red	dark red	cream	pink
Grain taste	v. good	poor	good	good	average	good	average
Grain storability	good	good	good	good	good	good	good
Brewing quality	v. good	good	good	good	good	good	good

[†] Data are based on farmers' perception of the performance of varieties, collected during two group interviews in Mkulula village. v = very.

interviews were carried out with farmers from different tribes. In-depth interviews were moreover done with farmers with special knowledge, such as an old woman in Ikuwala village who, as the only farmer in the area, maintained a large portfolio of finger millet varieties, or a rice farmer in Usangu, with a more stable supply of water than other farmers, who specialized in seed production. A third type of in-depth interview was with persons outside the communities who were important for farmers' plant genetic resources use, such as the manager of a local oil mill who provides modern varieties of sunflower seed to farmers.

- *Gender-specific interviews within individual households* were used to bring to light and clarify differences in the participation of the members of household in plant genetic resources management activities for different crops. In case a particular management function, such as seed selection of rice, was exclusively carried out by a particular sex (in this case women), this has obvious consequences for those who have knowledge about the management task.

Quantitative research methodologies

Stratified random sampling of 30 households to be included in a household questionnaire survey was done in each of the three villages. In none of the villages did a complete list of households exist and lists of households liable to taxation were used (these lists exclude the very old and the disabled). For practical reasons, the selection of farmers was not carried out totally at random, but by random selection from within accessible and representative subsections of individual villages. All the 90 selected households were willing to participate in the household survey.

A central hypothesis of the study was that management of plant genetic resources varies between different groups of people and three main factors of differentiation were assumed to be important: gender, ethnicity and wealth. While the actors are easy to identify for the first two factors, it is more complex to identify who is rich and who is poor in a village. Thus, a participatory *Wealth Ranking* was carried out in each of the three villages. In each village the wealth-ranking criteria were chosen by a small group of knowledgeable farmers following intense discussions. The wealth ranking criteria for the three villages are shown in Table 15.2.

The wealth-ranking criteria were applied on the group of farmers selected for the household questionnaire survey by the same group of knowledgeable farmers who had constructed them. Each household was discussed separately and there were rarely disagreements over which wealth category each particular household belonged to! From the discussion of each individual household, it became clear that the farmers' social character, in particular their work ethics, was an unwritten part of the farmers' holistic assessment of each household. In some cases, resource-poor but capable and hard-working young farmers were upgraded from the poor to the average category, while a non-resource-poor farmer in one case was downgraded to average as he was judged to spend too much time in the local beer club.

The household questionnaire was developed by the researcher in dialogue with farmers. The questionnaire was completed in the farmer's house and often both the husband and wife participated. The interview session lasted approximately 2 hours and the questionnaire included questions about: the household economy and assets, plant genetic resources management practices of all major crops cultivated, variety-specific crop production information on a field basis, dynamics of community seed exchange and genetic erosion, and farmers' perspectives and ideas on community-based solutions to problems of seed insecurity and poor access to new germplasm. With emphasis on capturing the dynamic aspects of plant genetic resources management, a follow-up survey was carried out among the same 90 farmers in 1999, 5 years after the original survey.

Table 15.2. Wealth-ranking criteria for Ikuwala, Mkulula and Nyelegete villages

Poor	Average	Non-poor
Ikuwala village		
-Limited ownership of poor-quality land only -Participate in reciprocal labour arrangements without being able to call for work parties on their own fields -Work as casual labourer for other farmers -Annual purchase of food from others	-May or may not own plough and a pair of draft oxen -May or may not hire casual labourers -May or may not own oxen or donkey-drawn cart -May or may not sell surplus crop production	-Ownership of a plough and one or more pair of draft oxen or a tractor -Access to large areas of quality land -Capacity to rent land for tomato cultivation in neighbouring village -Ownership of large herds of livestock (special for pastoralists) -Hire of casual labourers -Ownership of oxen or donkey-drawn cart -Annual sale of surplus crop production.
Mkulula village		
-Work as casual labourer for other farmers -Participate in reciprocal labour arrangements without being able to call for work parties on their own fields -Small cultivated area -Annual purchase of food from others	-Some own plough and a pair of draft oxen, others only a plough or only one oxen -Hire of casual labourers -Buy food in drought years and sell surplus crop production in better rainfall years	-Ownership of a plough and one or more pair of draft oxen -Hire of casual labourers -Ownership of oxen or donkey-drawn cart -Large cultivated area -Buy food in drought years and sell surplus crop production in better rainfall years
Nyelegete village		
-Less than 1.75 acres of rice under irrigation	-Production of between 20 and 60 bags of rice per year	-Production of more than 60 bags of rice per year

A quantitative statistical analysis of the household questionnaire data was carried out using the SPSS statistical programme. Farmers' holistic application of wealth-ranking criteria has been successfully used to analyze the data (Friis-Hansen 1999).

Reference

Friis-Hansen, E. 1999. The Socio-Economic Dynamics of Farmers' Local Plant Genetic Resources. A Framework for Analysis with Examples from a Tanzanian Case Study. CDR Working Paper 99.3. Copenhagen: Centre for Development Research. The publication can freely be downloaded from the internet: http://www.cdr.dk

16. The role of gender in the conservation, location and management of genetic diversity in potatoes, tarwi and maize in Pocoata, Bolivia

Lucio Iriarte, Litza Lazarte, Javier Franco, David Fernandez and Pablo Eyzaguirre[1]

Introduction

The indigenous communities of Bolivia are noted for the rich diversity of crops they use. Typically, a single community may manage 20 or more varieties of native potatoes along with a host of other traditional crop varieties of maize, quinoa and other grains and legumes as staple foods. These are maintained in order to adapt to the highly varied mountain environments and to cope with the marked stresses of cold, drought and disease that affect agriculture in the Bolivian Andes. The many varieties are also maintained to meet the multiple purposes and uses of these crops. This is evidenced in the large number of recipes and preparations which require specific local varieties and which are imbued with unique cultural values. The communities also maintain cultural systems that define distinct gender roles of cooperation and independent decision-making for men and women. Thus women and men have different ways and purposes for selecting crop varieties; they plant them in different niches within the complex mountain ecosystems in which they live.

This study examines differences in the ways that women and men manage the genetic resources of three traditional crops: native potatoes (*Solanum tuberosum*), tarwi (*Lupinus mutabilis*) and maize (*Zea mais*). This gender focus is applied in the context of ecosystem management. The study identifies the various niches and levels within the landscape to which men and women have common or differentiated access. Participatory approaches in collecting and using gender-disaggregated data on the management of local varieties are used to identify how gendered participation in the management of plant genetic resources supports the conservation of genetic diversity in crops.

The sociocultural, ecological and political setting

The participatory research using a gender-sensitive approach to assess the distribution and management of crop genetic diversity was organized by BIOSOMA, a Bolivian environmental NGO based in Cochabamba, Bolivia. The research involved four Quechua-speaking communities – Saqa-saqa, Cuchillera, Pocoata and Pila-pata – in the province of Arani in the Cochabamba region. This region is steeped in Inca history and Quechua culture. The local institutions survived the many centuries of *latifundio* where community lands were controlled by a few hispanized landowners. In 1953, the revolution and subsequent agrarian reform gave control of the lands back to the indigenous communities.

Pachamama, the earth mother in Quecha and Aymara cosmology, is the source of a unifying cult on managing land and agricultural resources including traditional crops. All agricultural

---

[1] This chapter is based on the 1999 publication in the joint IPGRI and FAO series, Gender and Genetic Resources Management, "El Rol del Género en la Conservación, Localización y Manejo de la Diversidad Genética de Papa, Tarwi y Maíz" by Lucuio Iriarte, Litza Lazarte, Javier Franco y David Fernandez. The authors of the original work are staff members of BIOSOMA in Cochabamba, Bolivia. Pablo Eyzaguirre of IPGRI prepared this English summary with the help of Carole Salas, also of IPGRI. For a complete report on the research results and the methods, the reader can consult the original publication.

activities are accompanied by rituals and ceremonies to repay Pachamama for the resources she provides. Pachamama reinforces the ecological relations between human communities and their natural environments. These rituals bring together men and women from households belonging to different clans and villages at various points in the agricultural cycle, such as field selection, planting and harvesting. Participation in the Pachamama cult is essential for participation in community-based PGR research.

Presently, the communities are organized as "*sindicatos*" to manage and defend their communal land and resource tenure. These local *sindicatos* are the primary institution where researchers can explain, negotiate and agree on a common set of research activities that can be shown to contribute to the social and economic development of the community. The deep attachment which local peoples have to their local crop varieties helped the BIOSOMA researchers to win support for research that would document and better understand the distribution and uses of local crop diversity and in particular how the decisions and actions of men and women can help to secure and improve the communities' vital biological assets. Furthermore, BIOSOMA had previously been contracted by the local *sindicatos* to deliver low-cost technical inputs for environmentally sound development of community agriculture. BIOSOMA's successful involvement in development actions to improve local welfare was one factor that underpinned the trust and enthusiastic participation of the local communities.

In sum, there are three fundamental lessons for the initial stage of participatory and gender-sensitive research on genetic resources. First, where possible it is good to combine development and research activities in a single programme. Second, identify local institutions that already have the authority and a claim on the participation of community members and integrate the participatory research into existing institutions and cultural values. Third, work to improve an existing resource that people already manage and "own". The research should strengthen their sense of ownership, and increase the value and options for the use of a community's existing biological assets, e.g. plant genetic resources.

In traditional communities such as those in Pocoata, men and women manage the same crop in different ways. This can be due to the different uses that men and women may have for the same crop. For example men may be more interested in the grain, women in the stover and craft materials. Or women and men may have access to different niches within the ecosystems at different times in the agricultural cycle. Finally, some crop varieties are considered primarily a women's crop or a men's crop. Out of these many options a range of diverse practices and uses have arisen that affect the distribution and range of genetic diversity in crops. In order to understand how PGR management practices vary according to gender, participatory approaches were used to document how communities classify and manage their agro-ecosystems and how they deploy plant genetic resources across the zones and niches within their landscape.

The communities of Pocoata identify three distinct zones within their agro-ecosystem:
- the highland zone (*puna*) at 3500–4100 meters above sea level (m asl)
- the intermediate slopes at 2900–3500 m asl
- the valley bottoms at 2700–29 000 m asl.

Participatory transects and mapping techniques permitted the community to identify the distinctive niches and the distribution of plant varieties across those niches. Women and men were interviewed together and separately in order to arrive at a final understanding of the complex environments in which farmers maintained and deployed their plant genetic resources. Gender-differentiated management of crop genetic resources is also seen in the temporal distribution of activities across the agricultural cycle. These differences were documented and are illustrated in Tables 16.1 to 16.3.

Table 16.1. Analysis of agricultural production calendar of the farmer communities

Activities	Gender†	J	F	M	A	M	J	J	A	S	O	N	D
CORN SEEDING													
Mishka		–	–	–	–	✗	✗	✗	–	–	–	–	–
Pocoata	MA, FA	–	–	–	–	✗	✗	–	–	–	–	–	–
Cuchillera	MA, FA, MC	–	–	–	–	✗	✗	–	–	–	–	–	–
Saqa saqa and Pila pata	MA, FA, MC	–	–	–	–	–	✗	✗	–	–	–	–	–
HIGH OR BIG SEEDING		–	–	–	–	–	–	–	–	✗	✗	✗	–
Pocoata	MA, FA, MC	–	–	–	–	–	–	–	–	–	✗	✗	–
Cuchillera	MA, FA, MC	–	–	–	–	–	–	–	–	–	✗	✗	–
Saqa saqa and Pila pata	MA, FA, MC	–	–	–	–	–	–	–	–	✗	✗	✗	–
CORN SEEDING		–	–	–	–	–	–	–	–	✗	✗	✗	✗
Pocoata	MA, FA	–	–	–	–	–	–	–	–	–	–	✗	✗
Cuchillera	MA, FA	–	–	–	–	–	–	–	–	✗	✗	–	–
TARWI SEEDING		–	–	–	–	–	–	–	–	–	✗	✗	✗
Cuchillera	MA, FA	–	–	–	–	–	–	–	–	–	✗	✗	✗
Saqa saqa and Pila pata	MA, FA	–	–	–	–	–	–	–	–	–	✗	✗	✗
CORN HARVESTING													
Chocio	FA, MC	–	✗	✗	–	–	–	–	–	–	–	–	–
Grain	MA, FA, MC	–	–	–	✗	✗	✗	–	–	–	–	–	–
POTATO HARVESTING													
Mishka	MA, FA	–	–	–	–	–	–	–	–	–	–	✗	✗
High seeding	MA, FA	–	–	✗	✗	✗	–	–	–	–	–	–	–
PROCESSING													
Potato (papa chuño)	MA, FA	–	–	–	–	–	–	✗	✗	–	–	–	–
Corn (wiñapu)	FA	–	–	–	–	–	✗	✗	✗	✗	✗	✗	✗
Tarwi (chuchus mote)	FA	✗	✗	✗	–	–	✗	✗	✗	✗	✗	✗	✗
Threshing of legumes	FA	–	–	–	–	–	✗	✗	✗	–	–	–	–
Grazing		✗	✗	✗	✗	✗	✗	✗	✗	✗	✗	✗	✗
Arid area		✗	✗	✗	✗	✗	–	–	✗	✗	✗	✗	✗
Cuchillera	FA, FC	✗	✗	✗	✗	✗	–	–	✗	✗	✗	✗	✗
Saqa saqa and Pila pata	FA, FC	✗	✗	✗	✗	✗	–	–	✗	✗	✗	✗	✗
Irrigated area		–	–	–	–	–	✗	✗	✗	✗	–	–	–
Cuchillera	FA, FC	–	–	–	–	–	✗	✗	✗	✗	–	–	–
Saqa saqa and Pila pata	FA, FC	–	–	–	–	–	✗	✗	✗	✗	–	–	–

† MA = Male adult; FA = Female adult; MC = Male child, FC = female child.

Methods

BIOSOMA in its work with the Pocoata communities applied a conceptual approach of "understanding and appreciation of farmers' logic and rationale in order to acquire a better scientific understanding of the processes that underpin the sustainable use of agrobiodiversity". This approach includes a variety of participatory tools and protocols (Fig. 16.1).

The first step was participation in the culture and daily life of the community. This included participation in community meetings, rituals and collective agricultural duties. The use of Quechua (the local language) as the medium of communication in the research also facilitated the eventual participation of the community. Men tend to be bilingual more often than women. By conducting the research in Quechua the researchers were assured of more balanced gender participation.

After gaining the trust of the community and familiarity with local values and customs that must be respected, the researchers may begin to discuss the scope and objectives of the research in order to develop a "letter of agreement" which stipulates the shared objectives and the scope of the participatory research. This agreement is the basis for community meetings on the

Table 16.2. Preference given to potato varieties according to the family farmers and the role of gender in the selection of varieties

Varieties	Number of persons that seed the variety	Who decides which variety to seed? †		
		Husband	Wife	Both
Imilla blanca	5	20.00	32.5	47.5
Waych'a	11	9.09	9.09	81.82
Ch'irawi	3	0.00	0.00	100.00
Waca loron ‡	6	0.00	50.00	50.00
C–óndor Imilla	2	0.00	0.00	100.00
Puka ñawi §	13	15.38	15.38	69.24
Khuchi aca ‡	1	0.00	0.00	100.00
Sani imilla	1	0.00	0.00	100.00
Khosi §	1	0.00	0.00	100.00
Saiciri	1	0.00	0.00	100.00
Huaman ‡	1	0.00	100.00	0.00
Amajía ‡	1	0.00	100.00	0.00
Kunurana ‡	1	0.00	100.00	0.00
Runa toralapa §	1	10.00	60.00	30.00
Ignacio papa	2	0.00	0.00	100.00
Apasana	8	12.50	50.00	37.50
Iqari	10	20.00	50.00	30.00
Puka huayq'u ‡	7	14.29	57.14	28.57
Sani runa	4	0.00	100.00	0.00
Lungha	1	10.50	75.50	14.00
Llogalla papa ‡	2	0.00	80.00	20.00
Condorillo	2	25.00	50.00	25.00
Runa	7	0.00	71.43	28.57
Luck'y	2	0.00	50.00	50.00
Ñucch'a papa ‡	1	0.00	100.00	0.00
Llust'a papa	6	33.33	50.00	16.67

† Percentage of decisions according to varieties.   ‡ = Huayq'u papas o papa huayq'us.
§ = Improved varieties.

Table 16.3. Distribution of activities according to gender during the harvest of tarwi. Primary responsibility for activity is shown in **bold**.

Activity	Gender†	
	Intermediate area	Highland area
In the seeding phase		
Opening of furrows	MA, FA, CH	MA, FA, CH
Planting of seeds	**MA,** FA, CH	**MA,** FA, CH
In the post-harvest phase		
Cutting of plants	**MA,** FA, CH	**MA,** FA, CH
Harvest of leguminous vegetables	**MA,** FA, CH	**MA,** FA, CH
Threshing	**MA,** FA, CH	**MA,** FA, CH
Drying	**MA,** FA, CH	**MA,** FA, CH
Harvesting of grain	**MA,** FA, CH	**MA,** FA, CH
Bagging	**MA,** FA, CH	**MA,** FA, CH
Marketing	**MA,** FA, CH	**MA,** FA, CH

† MA = Male adult; FA = Female adult; CH = Child.

priorities and problems which the research should address. During this process the various research techniques to be used are discussed so that community members can determine the level and nature of their participation.

The next step in the research was to survey the landscape managed by the communities. This involved the characterization of the different zones and niches using the local categories for distinguishing these spaces (Fig. 16.2). Researchers conducted transect walks with members of the community and sketched these transects to discuss how women and men deploy plant genetic resources across these spaces. As the research was focused on three species – maize, tarwi and potato – it was possible to begin immediately identifying the range and distribution of the different growing environments and to see how they differed between men and women.

The fourth step was to conduct informal extended interviews with families and to identify key informants in the process. This was a crucial step in understanding gender-differentiated management of PGR because it enables researchers to understand how decisions are made within the household and the community. In some cases decisions were made jointly by men and women, in other cases decisions were gender specific. The interviews also gave indications of where different decisions according to gender created a wider and more varied deployment and use of germplasm for traditional varieties of maize, potato and tarwi.

Once the researchers had acquired sufficient insights into community and household dynamics and decision-making, a questionnaire was developed. The questionnaires were used to compile data on actual practices and choices of material according to gender, crop variety and zone within the landscape. When applying participatory approaches, the questionnaire is

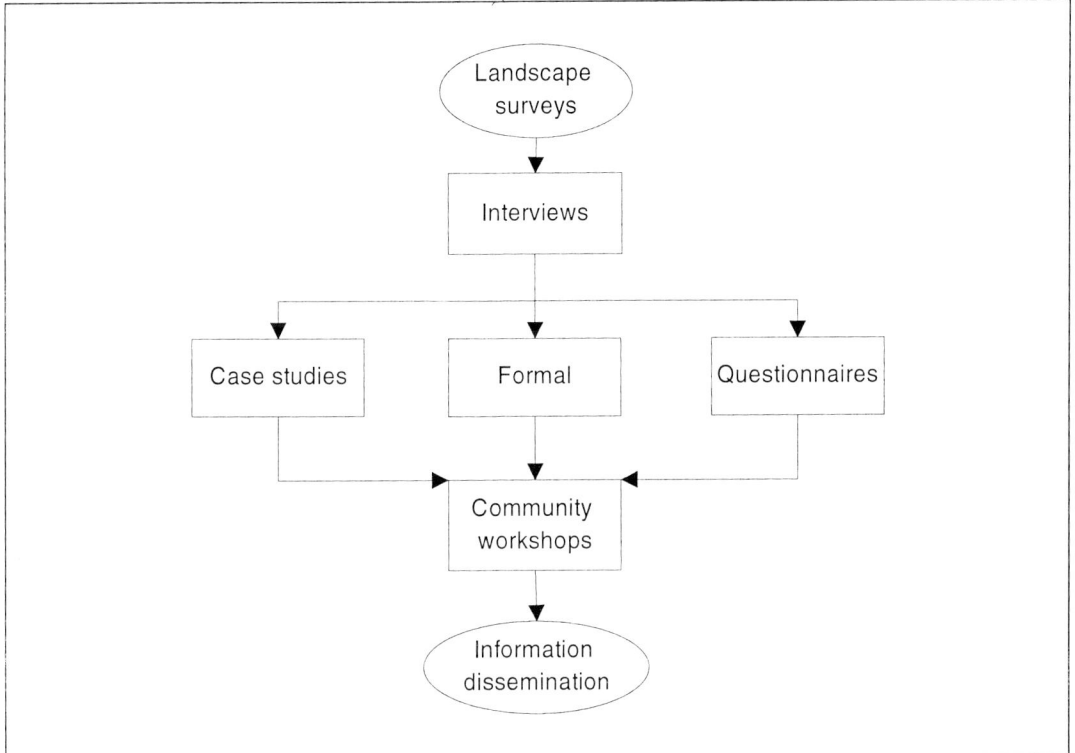

Fig. 16.1. Schematic of the participatory research method.

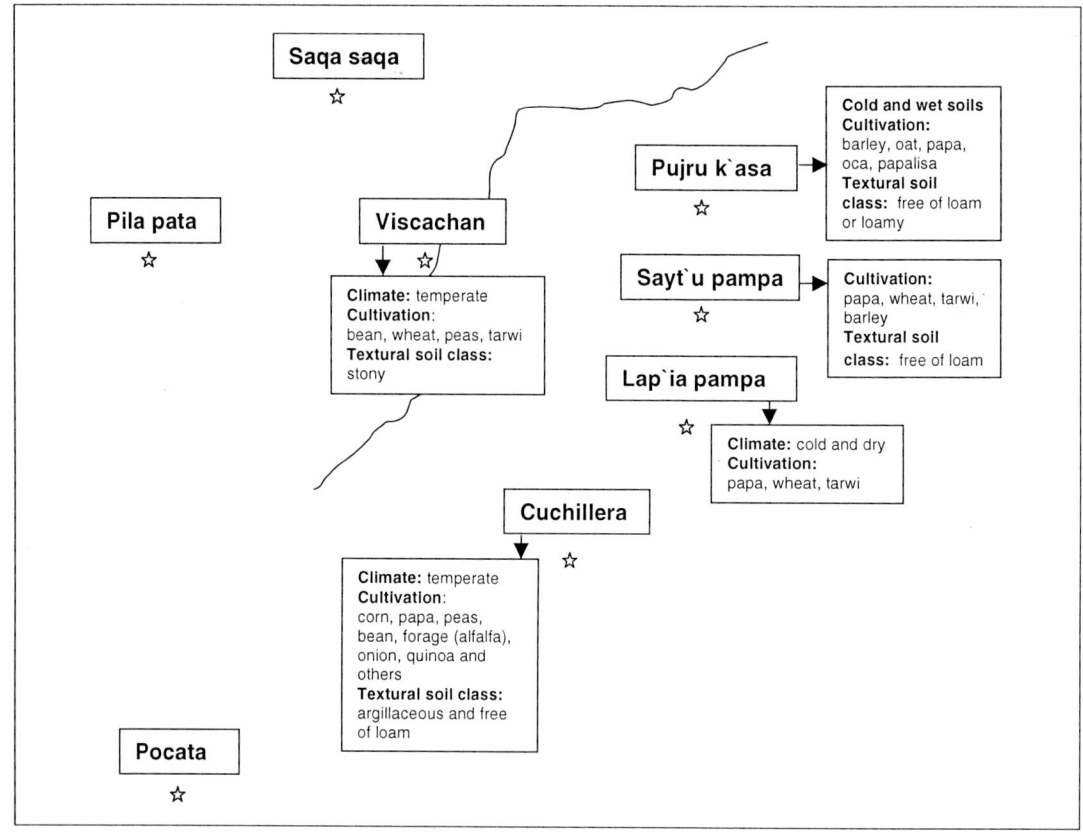

Fig. 16.2. Description of the agro-ecosystem of Cuchillera, as defined by the farmers of the area.

a supplemental tool used in the later stages of the research. In standard social science research with agrarian communities, the questionnaire is often the first step and the only point of interaction between the researcher and the community. The questionnaire serves as a very effective barrier insulating the researcher from the acquisition of any knowledge or wisdom that farmers might have. At the same time it consumes a great deal of the scarce time of farmers without any visible benefit in return. Thus the traditional use of questionnaires effectively keeps researchers and farming communities apart and may be an important reason why some researchers rely on it. As a development NGO with long-term commitment to the region, BIOSOMA's use of questionnaires was judicious and complemented other more dynamic and participatory techniques.

Case studies were conducted on a small sample of households. The case studies focused in detail on the decision-making and management practices with the stock of potato, maize and tarwi germplasm that each household managed. The dynamics of decision-making and allocation of responsibilities and tasks were also further assessed.

Formal interviews were conducted with selected key informants on particular topics such as seed storage, management of varieties for stress resistance, identification of differences (if any) in the management of the same varieties. Formal interviews were also used to elicit community perceptions of changes in the use and deployment of plant genetic resources over time. The formal interviews tended to focus more on women as their views may tend to be less prominently expressed in collective or public interviews.

Once the key parameters for the conservation, location, identification and management of potato, maize and tarwi genetic resources had been identified, a community workshop was organized with the help of community members who were active in the research and were literate. The goal of the workshop was to present and validate the results of the research with the community. Community members participated actively and gained a better understanding of the process by which they manage their landscape, their genetic resources, and how their culture and in particular gender plays a part in maintaining the necessary diversity to survive in the fragile conditions of the Andes. Community members also were able to identify points in their management systems where improvements could be made in techniques or in acquiring training of women and men.

Conclusion

By focusing on gender and genetic resources management in the traditional crops that are most valued in the community, a development partnership was forged based on improvement of the existing and valued biological assets of the community. As a biodiversity and development organization BIOSOMA was able to continue working with the acquired farmer knowledge to improve the management of the agro-ecosystems in Pocoata. Participatory PGR approaches empower local communities to maintain their resource base and their rights to their unique culture and environment. The study in Pocoata showed that men and women had distinctive roles in managing genetic resources. Women were more likely to be the key decision-makers on seed selection and storage (PGR conservation), while men were more likely to decide where these varieties were to be deployed (PGR deployment and location) within the landscape. Most important was a conscious sharing of responsibilities across genders that characterized the system of PGR management in Pocoata. Any exclusion of one or the other gender would undermine the social and cultural fabric that underpins agrobiodiversity in the region. The cultural practices, decision-making and the need to adapt varieties to the local environment continue to provide a competitive edge to local cultivars of maize, potato and tarwi. The modern cultivars, while conferring commercial gain primarily to men, have placed households at risk where their food security is concerned. One of the recommendations is to continue to work on the local varieties identified in this study (see Table 16.2 for example of native potato varieties) to ensure that they enhance the quality and availability of the germplasm.

Section IV

Participatory approaches for establishing community seed banks and improving local seed systems

17. Overview

Esbern Friis-Hansen and Bhuwon Sthapit

A dual system of seed supply exists in most developing countries, with (1) a formal system of organized seed production, either privately or publicly owned, producing and marketing modern varieties and (2) an informal system based on local farmers' retention of seed from previous harvests, storage, treatment and exchange of this seed within and between communities. Public sector provision of improved seeds in developing countries has often failed because of inefficiency of government. As part of structural adjustment, public seed industries recently have been targeted for privatization and regulatory reform (Tripp *et al.* 1997). However, the private seed industry cannot meet the seed demands of small-scale farmers, who typically desire multiple varieties of seed for all crops which may vary from one season to the next, in small amounts, at the right time and at a reasonable cost. The very high transaction costs of delivering a range of appropriate seed to marginal areas makes it unprofitable for the commercial seed industry to play a major role (Wiggins and Cromwell 1995). Farmers' own informal seed systems have remained the dominant source of seed in most developing countries, in particular for food crops. The informal seed sector is typically based on indigenous structures for information flow and exchange of seed. It operates on a small scale at the community level (Cromwell *et al.* 1992). In spite of its importance and obvious possibilities for improvement, the informal seed sector has until now received little attention or support from the scientific community or the state.

Activity 13 of The Global Plan of Action (FAO 1996) identifies a need to strengthen local capacity to produce and distribute seed of many crop varieties, including landraces, that are useful for diverse and evolving farming systems. The GPA goes on to emphasize a focus on varietal needs of resource-poor farmers and on crops that are inadequately covered by the private sector. Participatory activities to address small-scale farmers' inadequate and insecure seed supply in the 1990s have focused on two areas: (1) decentralized multiplication of modern varieties which are not produced by the formal seed industry and (2) organizational and scientific support of community seed exchange and storage systems.

Evidence from a review of NGO programmes involved with decentralized multiplication of seed suggests that such programmes have not been participatory in nature, that they operate on a small scale and lack technical competence, and that they are facing serious problems of economic viability (Wiggins and Cromwell 1995). The review suggests supporting the growth of small seed production and distribution enterprises at farm and village levels. This section will focus on participatory approaches of improving the community seed exchange and storage systems, in particular through community seed banks and by improving the physiological quality of local landraces. The main objectives of community genebanks in Ethiopia are: (1) to serve as a decentralized *ex situ* facility that is a reserve of locally used seed for farmers, and (2) to facilitate access for farmers to crop genetic diversity from the national genebank, in particular so-called 'elite landraces'.

Even in diversity-rich areas farmers cannot maintain a large number of cultivars at their own farm. Usually farmers or even NGOs do not have easy access to the genetic materials they have donated. Exchange of seed and introduction of new seed from local markets, seed fairs or other informal-system components is an important way of conserving a large amount of genetic diversity within a small area. Community biodiversity registers and community seed banks can address this vital need which cannot be met by the centralized national seed system or genebank.

Chapter 18 discusses the experience of community genebanks as a strategy to mitigate the effects of genetic erosion of durum wheat landraces in the Addaa region of Ethiopia. With a dual strategy of providing a secure seed reserve of local varieties, while promoting enhanced landraces from elsewhere, the community genebank programme has created a credible alternative way of increasing productivity for low external input farmers.

Chapter 19 discusses organizational support for local knowledge and institutions for plant genetic resources management of rice and other crops in Bangladesh. The activities are part of a movement for ecological agriculture which aims at improving agricultural production without use of external chemical inputs. The approach emphasizes the creation of organized seed networks between member farmers with the responsibility of ensuring both *in situ* conservation of agrobiodiversity and *ex situ* conservation at the household and community levels. This approach is novel as it seeks to formalize and coordinate what individual farmers are already doing as part of their plant genetic resources management. These household-based seed networks are linked to community seed wealth centres, which function as community seed banks and sources of access to varieties of landraces from outside the community.

Planting material for small-scale farmers of roots and tubers is rarely supplied by the formal seed industry and there are only a few cases where the farmers' indigenous systems of seed supply are receiving outside support. One such example is discussed in Chapter 20, which reviews the experience of scientific support for improving landraces by eliminating viruses of local potato varieties and returning virus-free potato seed to farmers and communities which are specialized in seed production. The programme has been able to double yields for farmers and the participatory approach used has the potential to be adopted successfully elsewhere.

References

Cromwell, E., E. Friis-Hansen and M. Turner. 1992. The Organisation of the Seed Sector in Developing Countries, a Conceptual Framework of Analysis, pp. 1-81. ODI, London.

FAO. 1996. Global Plan of Action for the Conservation and Sustainable Utilisation of Plant Genetic Resources for Food and Agriculture and the Leipzig Declaration, pp. 3-63. International Technical Conference on Plant Genetic Resources, FAO, Rome.

Tripp, R., N. Louwaars, W. Joost van der Burg, D.S. Virk and J.R. Witcombe. 1997. Alternatives for Seed Regulatory Reform, an Analysis of Variety Testing, Variety Regulation and Seed Quality Control, pp. 1-25. Agricultural Research & Extension Service, ODI, London.

Wiggins, S. and E. Cromwell. 1995. NGOs and seed provision to smallholders in developing countries. World Development 23(3):413-422.

18. Community seed banks and seed exchange in Ethiopia: a farmer-led approach

Regassa Feyissa

Introduction

The seed system under most farming systems in Ethiopia is characterized by the local multiplication of seeds by farmers themselves. Farmers practise seed selection, production, saving of planting materials and exchange of seeds within and among the farming communities. Seed production in most cases is non-specialized, but is integrated into the production of grains, roots and tubers for consumption and marketing. This traditional seed-supply system is an important back-up to overall agricultural crop production in the country. It is mainly based on the farmers' varieties with the exception of cases where the seed system depends on improved or introduced crop varieties. Usually, the dependency on introduced varieties is due to displacement of farmers' own varieties.

Farmers' objectives and needs vary according to the local seed system, and this determines the strategy of external support for on-farm management of crop diversity. In cases where crop genetic diversity still exists on farms, for example, the major task of external support or on-farm crop management activities is to develop competitive forms of farmers' varieties in order to promote production and ensure conservation through continued use. It is understood that such a production promotion approach may entail risks of genetic erosion unless careful and systematic conservation measures are taken. Where genetic diversity is threatened and farmers' options are shrinking, the main objective is to replace the lost diversity in order to maximize options at that level. This is the case in Addaa region, where the local community seed bank system serves as an instrument for this objective.

Background for developing a community seed bank system in Addaa region

Addaa region is one of the high production potential areas for cereals in Ethiopia and was once known for a wide range of diversity in durum wheat. It is an area where displacement of durum wheat due to intensive introduction of improved bread wheat varieties at present has reached about 95% (Feyissa 1997). Sorghum and oil crops, which were once widely grown in the area, are hardly cultivated any more. Owing to a high market price for tef (*Eragrostis tef*) and wide distribution of improved bread wheat, these two crops today dominate the cropping system in the region. Faba bean and field peas are also threatened by displacement, diseases and pests. The use of traditional agronomic practices associated with farmers' cultivation of a diversity of crop species – such as fallowing and crop rotation – is also declining and the knowledge about them is at risk of being lost. Nevertheless, some farmers in Addaa have been reluctant to discard the old varieties and maintain them through storage or cultivation in small fields or patches.

The on-farm crop diversity management programme in the Addaa region as a whole was started in 1996 with the involvement of about 25 farmers at the initial stage. It is financially supported through the Global Environmental Facility (GEF) project fund for farmer-based conservation of plant genetic resources in Ethiopia. Prior to that, on-farm crop management initiatives Seeds of Survival (SoS) programme in the region, supported by the Unitarian Service Committee (USC) of Canada, did a great deal in creating awareness about the value of genetic diversity. As a result, the number of voluntarily involved farmers in the programme increased to 157, 385 and 579 in the 1997, 1998 and 1999 cropping seasons, respectively.

The USC/Canada helps support a SoS programme in which materials collected from nearby areas are given to farmers to plant and to carry out simple mass selection to improve their characteristics. Access to a diversity of crop varieties is an essential element for achieving sustainable agricultural development in the region. The strategy of the programme, therefore, was to initiate on-farm crop conservation and simultaneously seek ways to increase productivity through enhancing farmers' access to crop diversity. As farmers' goals and needs were central in the programme, materials for on-farm crop genetic diversity management included all available farmers' varieties and in some cases formally modified forms of these varieties.

Compared with other parts of Ethiopia, the Addaa region has the highest use of external agricultural inputs, i.e. fertilizers and pesticides. Farmers have therefore become highly dependent on market conditions and are vulnerable to the recent increase in fertilizer cost, which has not been matched with an equal increase in farm gate crop prices. Involvement of farmers in safeguarding the existing local crop diversity through *in situ* conservation in the region requires incentives for farmers. The incentives are such that they offer farmers competitive advantages with the existing high input/high output cropping system, which is based on cultivation of modern varieties in monoculutre. As an instrument to create competitive advantages for farmers through external support, a community seed bank system was developed in two strategic localities (Ejere and Cheffedonsa) in the region.

One of the major objectives of the external support was to restore the disrupted local systems through the creation of new opportunities for farmers at the farming community level. It was also to reintroduce the forgotten aspects of traditional practices and knowledge of crop conservation and variety development. The need for the latter was confirmed in one of the efforts made to restore some of the durum wheat germplasm accessions collected 25 years ago from the region. According to the observations made, it was realized that farmers below the age of 35, who are the majority in the farming communities in the region, could hardly recognize those farmers' varieties. It was only the elders above 50 who were able to recognize and identify the varieties and the agronomic practices they require (Feyissa 1998). This resulted from the long-time detachment of farmers from their varieties, and consequently suggests the need for farmer participation and for a joint venture of farmers and scientists in the process of establishing an on-farm crop management scheme in the region.

A community genebank is a system under which community-based crop diversity management is practised. Restoration of diversity and strengthening of traditionally employed farmers' crop selection practices, as well free flow of seed through seed exchange schemes, are among the major elements of the system. Such an approach in the region has offered the opportunity for integrating formal and informal efforts in order to link conservation to production at community level (Feyissa 1999).

The community seed bank system was created to operate not solely as an income-generating set-up that distributes seeds on a loan basis. Rather, it was initiated with the following objectives: (1) to serve as a community-based *ex situ* facility for seed reserve and as a grain repository, (2) to satisfy farmers' needs by creating access to crop genetic diversity, (3) to ensure a sustainable supply of planting materials for farmers, and where possible (4) to facilitate proper market access for their produce and the materials they require. At the moment, the two community seed banks hold over 10 t of 18 elite wheat varieties and 12 mother populations of durum wheat used for *in situ* conservation; and over 2 t of seeds of different legumes. Duplicates of the same varieties are found in the possession of most of the programme member farmers.

A participatory approach to the management of community seed banks in Addaa region

Respect for and identification of farmers' needs and objectives are essential when establishing a comprehensive system for crop genetic resources management on-farm. It was on this principle that a community genebank system, which includes on-farm conservation and community seed banks, was established in the Addaa region. The community seed banks operate as community-based seed networks and seed-supply systems for locally adapted crops and enhanced farmers' varieties. The aim of the system is to ensure sustainable supply of planting materials for farmers for both conservation and production purposes. Since its establishment the system has helped to minimize farmers' dependency on high-input varieties and has enabled facilitation of better market access for farmers' produce.

The combination of access to a diversity of improved and low-input farmers' varieties, and better access to crop market, has provided adequate incentives for farmers to be involved in the maintenance and development of crop diversity on-farm. The incentives include both market and non-market ones. The non-market incentive includes farmers' access to the planting materials of interest to them, and seed security. Seed security was demonstrated in the 1997 cropping season when farmers lost crops due to excess rain, and were able to get seeds for the 1998 cropping season from the seed banks. This situation has attracted farmers wishing to be involved in the programme. As a result, over the last 4 years, the occurrence of durum wheat landraces has increased by 38% at Ejere locality in Addaa region (Feyissa 1998).

To date, both *in situ* conservation and seed bank systems are managed by the Crop Conservation Associations (CCAs), which were established as representative bodies of the Farmers' Associations in the programme sites. The Farmers' Associations are the legally established bodies according to the current Agricultural Development Policy of Ethiopia. Rural development activities of various nature are usually linked to these organized bodies. For strategic reasons, the community seed banks were also established as part of the rural community development activities, and operate as central seed and grain repositories at this level. They are linked to the National Genebank and to the village-level local storage facilities existing within the Farmers' Associations. Usually the village-level farmers' storage facilities maintain varieties used for various purposes by farmers themselves. These varieties are duplicated in the seed banks and in the National Genebank.

The curators of the CCAs are elderly women and men who fully decide on the management of the activities of the Associations. Involvement of women in the management and as beneficiaries has increased to 21%. Under the CCAs, trained farmer-trainer groups are actively engaged in the management and monitoring of the seed bank, in seed distribution and collecting, in seed marketing and in other activities of on-farm crop management. Selection is usually conducted by older farmers, both women and men, and with scientists. The scientists from the Ethiopian Biodiversity Institute and agricultural development agents working in the programme sites assist the CCAs.

The community seed bank system has offered the opportunity to directly link the formal genebank to the community-based on-farm crop management activities. On the basis of studies made to identify factors that sustain on-farm conservation under the crop production system in the Addaa region, a model for linking formal *ex situ* to farmer-based *ex situ* and *in situ* activities has been developed (Fig. 18.1). As illustrated in the model, the National Genebank supplies the CCAs with germplasm materials and facilitates capacity-building of the system through financial support from the GEF project fund. The National Genebank technically supports the whole system, including the training of farmers and extension agents involved in the activities.

The community seed banks offer various community services such as seed security, seed distribution and exchange, germplasm restoration and introduction, seed marketing and other services as needed.

Seed marketing is done through a revolving fund allotted for the seed banks. In some cases, this mechanism has helped to protect farmers from being victims of low prices (usually fixed by other groups such as grain merchants). Access to proper market price for their produce, local seed security and availability of varieties of their interest have therefore become strong incentives for farmers' involvement in on-farm conservation and development of crop diversity on a voluntary basis.

The model also has enabled a collaboration of farmers, extension agents and researchers as illustrated in Figure 18.2. In the process, farmer's traditional agronomic practices, complemented with modern technologies, are strengthened. This includes soil conservation and crop rotation practices as well as pest management, which have become additional attractive incentives for farmers. Integration of efforts and activities in this manner has helped to avoid or minimize the conflicts between modernization and introduction of new technologies, and management of crop genetic diversity on-farm.

The approach also has enabled the complementarity between the formal and the informal efforts, providing an opportunity to fill the exiting gaps between the use of high-input varieties and the need to increase crop production through the maintenance and enhancement of better-adapted farmers' varieties. It is envisaged that the experience gained at Addaa will be expanded into other farming systems in the country, but with modifications as deemed appropriate. The long-term sustainability of the developed system depends on the consistency of commitments of all kinds to further support the system. Involvement of alternative or supportive extension channels such as non-governmental organizations would definitely contribute to the expansion and strengthening of the system. At the moment the effect of the programme in general has attracted the attention of the decision-makers, especially of those who are deciding on the agricultural development directions in the region.

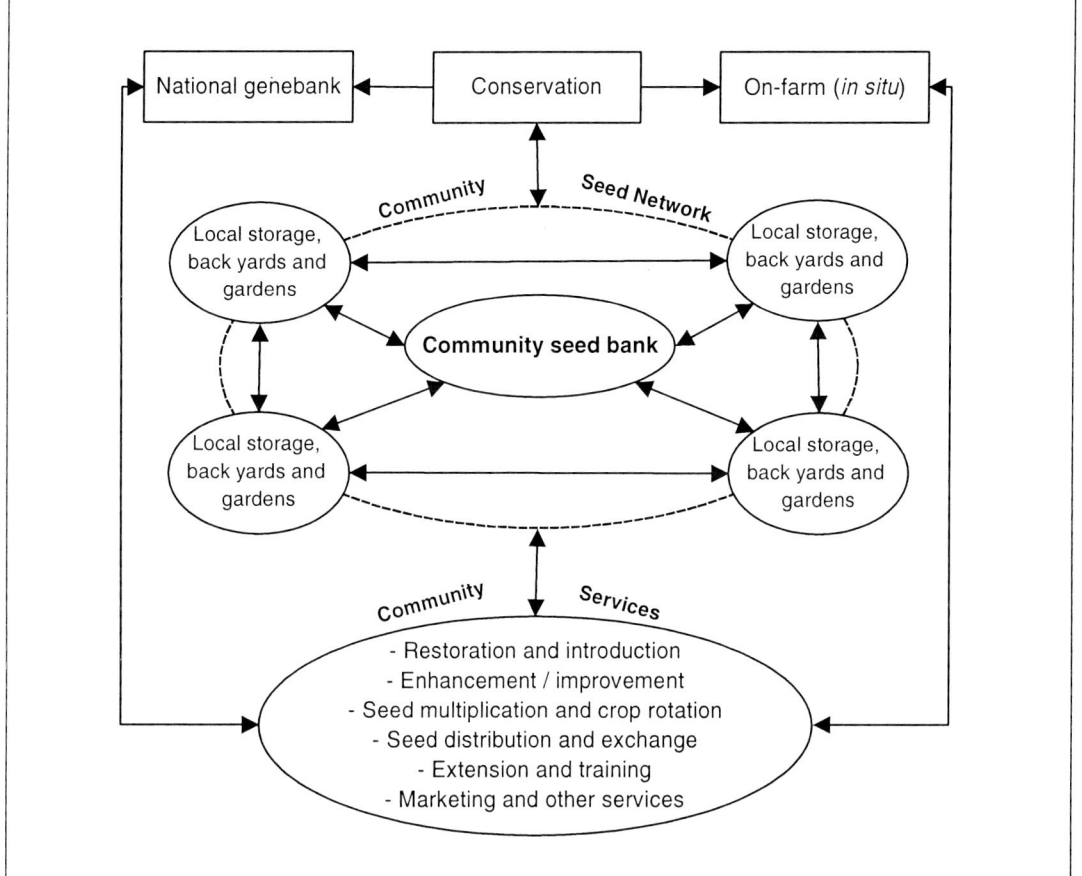

Fig. 18.1. Operation of the Community Seed Bank System in Addaa region, Ethiopia.

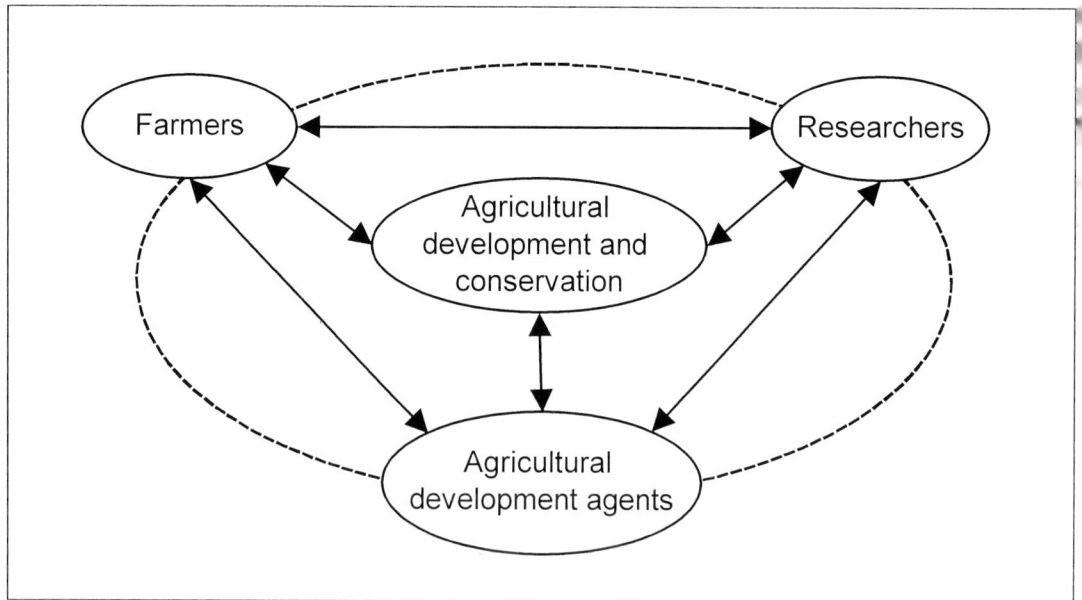

Fig. 18.2. Integration of efforts of the major stakeholders.

The impact of the programme

Promotion of sustainable capacity of the farming communities in food self-sufficiency needs to be one among the major objectives of a community-based genetic resource management programme. The strategies of the management should contribute to the provision of alternative ways of increasing productivity without additional cost to the farmers. It should also be implemented through the empowerment of the local communities and through integrated efforts to enhance livelihood security at that level. External support for on-farm conservation should ensure local farmers social and economic benefits accruing from the maintenance of crop genetic diversity through diversity-based production.

In the Addaa region, introduction of enhanced low-input farmers' varieties and reintroduction of some of the displaced legume crops through the community seed bank activities strongly attracted farmers to be involved in the crop conservation association. Over the last 3 years, the number of farmers using enhanced farmers' varieties of wheat together with local legume crops for crop rotation has increased by 160%. There are three important reasons for this: (1) pulses are essential components of the food system at the rural community level, and most of them are scarce, (2) the relative price of pulses at the moment is high, (3) the nitrogen-fixing effect of pulses cultivated in crop rotation reduces farmers' dependence on expensive chemical fertilizers. In effect, economic attractiveness is the best incentive for re-introducing agricultural diversity.

Using landraces in their competitive forms is an attractive alternative for farmers since they perform well under low-input farming conditions (Worede 1992; Worede and Mekbib 1993). In particular, rotating cereals with legumes instead of using chemical fertilizers on marginal lands is more beneficial to farmers because of its cost-effectiveness. That is why access to legume crops has become one of the important incentives for farmers with limited options. In one case in Addaa region, for example, a farmer using manure together with crop rotation has increased the yield of enhanced farmers' variety to 4 t/ha, while the average yield of an improved high-input variety in the same area could not exceed 2.2 t/ha.

According to a study conducted in the 1998-99 cropping season, growing enhanced farmers' varieties of wheat through crop rotation is more cost-effective and beneficial to subsistence farmers than growing improved wheat varieties using chemical fertilizers. In Ejere locality, for example, the average yield for improved high-input varieties (IHIV), and low-input farmers' varieties (LIFV) of wheat was 1700 and 1608 kg/ha respectively. The yield level range was between 1000 and 2500 kg/ha for both IHIV and LIFV, while IHIV is sensitive to agro-ecological and management conditions. The average market price of the year for both LIFV and IHIV was the same, 195 Bir/100 kg (approx. US$24/100 kg). When IHIV was grown, farmer's grain sale was 3315 Bir/ha. The cost of the chemical input for the year was 338 Bir/ha, and this brought the net gain of a farmer down to 2977 Bir/ha. When LIFV was grown using crop rotation, the grain sale was 3136 Bir/ha. Since the farmer used crop rotation, the cost of input was not incurred and the net gain of the farmer remained 3136 Bir/ha (Table 18.1). Thus, when the cost of production of the two varieties is compared, it is advantageous for a subsistence farmer to grow LIFV using crop rotation rather than IHIV using chemical fertilizer.

There is no question that with intensive management and input use, properly developed improved varieties give high yield. What is of concern here is whether marginalized farmers can make such intensive management and investment in order to achieve a positive output, which by itself is an incentive for increased production. Since the use of chemical inputs is an additional cost to subsistence farmers, it can only be beneficial if the output pays the farmer's production cost.

In most cases, subsistence farmers do not have access to a proper market, or can hardly determine prices for their produce. The price is determined by externalities such as grain merchants, imposition on input debt payment, emergency cases, etc. According to the study made in the same cropping season, farmers lose a significant portion of their net income because

Table 18.1. Cost-effectiveness of crop rotation to subsistence farmers

Crop type	Yield (t/ha) Range	Yield (t/ha) Avg.	Input use (kg/ha) Urea	Input use (kg/ha) DAP	Input cost/ha (Bir) Urea	Input cost/ha (Bir) DAP	Input cost/ha (Bir) Total	Crop price/ 100 kg	Total gain (Bir)	Production cost (Bir)	Net gain (Bir)
Tef	0.6-1.2	0.9	100	200	150	480	630	270	2430	653	1777
Wheat (IHIV)	1.0-2.5	1.7	50	100	75	240	315	195	3315	338	2977
Wheat (LIFV)	1000-2500	1608	Crop rotation					195	3136	–	3136

Table 18.2. The impact of lack of better market on farmers' gain

Crop type	Yield (kg/ha)	Price/100 kg Farm gate market	Price/100 kg Urban market	Gain/ha (Bir) Farm gate market	Gain/ha (Bir) Urban market	Differ. (Bir)	Expense to reach urban market	Net loss/ha (Bir)	% Net loss/ha
Tef	900	270	305	2430	2745	315	–	–	
Wheat	1700	195	220	3315	3740	425	184	241	6
Grasspea	2170	160	190	3472	4123	651	261	390	9
Chickpea	1020	180	210	1836	2142	306	122	186	9
Fieldpea	730	280	325	2044	2373	329	87	248	10
Fenugreek	660	300	350	1980	2310	330	79	264	11
Lentil	600	290	340	1740	2040	300	72	228	11

of lack of access to a proper market. For example, when a farmer sells his produce at farm gate price, he loses up to 11% of his net income per hectare (Table 18.2). For subsistence farmers, such level of gain/loss has a huge implication for family life. That is how by facilitating better market access for farmers' produce, the community seed bank system is working toward the improvement of farmers' seed security and socioeconomic gain.

Conclusion

The on-farm crop management system should be able to provide an alternative way of increasing productivity, and reducing farmers' dependency on the use of external inputs. As a lesson learned, farmers strictly consider the benefits they may gain from the conservation of diversity on-farm. They take into account the benefit of growing one or another crop, which in turn may depend on achievable yields, available technology and market incentives, as well as on cost of inputs and outputs. Farmers' interest in integrated pest management (IPM), for example, mainly lies in its lower cost and its effect on yield. Similarly, subsistence farmers do not maintain diversity just for the sake of conservation alone, as the value of conserving crop diversity lies in its use.

The major challenge in sustaining on-farm crop diversity management therefore lies in making it economically viable and self-supporting. Market and non-market incentives for farmers' produce, and integrated extension services that deal with agronomic, socioeconomic and cultural practices, are important elements to maintain the viability of the system. A seed bank as a system can only be sustained whenever it operates as an element of a community-based seed supply system, satisfies farmers' needs and objectives, and supports the livelihood security of the farming communities. The case in Addaa manifests this fact.

References

Feyissa, R. 1997. *In situ* conservation of food crops: an Ethiopian model. *In* Proceedings of the workshop on planning and priority setting in eco-geographic survey and ethnobotanical research in relation to genetic resources in Ethiopia, 15-16 Feb. 1997, Addis Abeba.

Feyissa, R. 1998. A Dynamic Farmer-Based Approach to the Conservation of Ethiopia's Plant Genetic Resources. Project Progress Report, Addis Abeba.

Feyissa, R. 1999. Mainstreaming Biodiversity Conservation towards Sustainable Agricultural Development: An Ethiopian Perspective (In press).

Worede, M. 1992. Ethiopia: A genebank working with farmers. *In* Growing Diversity. Genetic Resources and Local Food Security (D. Cooper, Renee Vellve and Henk Hobbelink, eds.). Intermediate Technology Publications, London.

Worede, M. and H. Mekbib. 1993. Linking genetic resources conservation to farmers in Ethiopia. Pp. 78-84 *in* Cultivating Knowledge: Genetic Diversity, Farmer Experimentation and Crop Research (Walter de Boef, Kojo Amanor, Kate Wellard and Anthony Bebbington, eds.). Intermediate Technology Publications, London.

19. Seed conservation and management: participatory approaches of *Nayakrishi* Seed Network in Bangladesh

Farhad Mazhar

Introduction

The *Nayakrishi Andolon* (New Agricultural Movement) of Bangladesh is a movement for ecological agriculture. It is based on simple principles like no use of pesticides and chemicals, soil management through recycling of local resources rather than using external input of fertilizers, and the practice of mixed cropping and crop rotation for pest management and reduction of risk due to crop failure. Mixed cropping is also crucial to improve soil productivity and increase crop biodiversity. The central approach of the initiative squarely lies in the conservation, management and use of local seed and genetic resources and adopting and improving production techniques suitable for farmers' seed. Thus hundreds of local varieties of rice, vegetables, fruit, timber trees and many other crops have been reintroduced within a short period of time. Farmers in the *Nayakrishi* area are estimated to cultivate at least 200 varieties of rice, and this number is increasing. The movement now is negotiating with the national genebank to help them regenerate the collected germplasm and internalize the conservation of genetic resources as an inbuilt operation of the movement.

The success of the approach is directly related to the commitment of the movement to evolve upon the local and indigenous knowledge systems and critically integrate the successes, failures and insights of modern science. The movement does not mechanically separate formal and informal knowledge systems, and does not subscribe to the two-system theory of knowledge. It strongly believes in the capacity of the farmers as authentic knowledge producer, no matter how it is articulated, orally or otherwise. At least 50 000 farmers all over Bangladesh practise *Nayakrishi*. A seed network of these farmers ensures conservation and use of local genetic resources found in the area. The organizational approach of the movement to the conservation of seed and genetic resources is described here.

The Nayakrishi Seed Network

The *Nayakrishi* Seed Network (NSN) is the active farmers' network within *Nayakrishi Andolon* with specific responsibility for ensuring both *in situ* conservation of biodiversity and genetic resources in the farming field and *ex situ* conservation at the household and community levels. The precise responsibility of the NSN, within the ecological food production practices and organizational activities of *Nayakrishi Andolon*, is to ensure collecting, conservation, distribution and enhancement of seed/germplasm among the members of *Nayakrishi Andolon*. The seed conservation activities take into account the particular ecological features of villages. It is the ecological features, therefore, that determine the clustering for NSN, not the administrative boundary of the village.

The NSN builds on the farming household, the nodal interactive point for *in situ* and *ex situ* conservation. Farmers maintain diversity in the field, but at the same time conserve seed in their homes for several years to be replanted in the coming seasons. The NSN is a network of *Nayakrishi* Community Seed Networks (NCSNs) established in different villages or areas. *Nayakrishi Andolon* constitutes NCSN from farmers who are experienced in seed collecting, quality maintenance, conservation and use. Sharing of seeds and germplasm among members of the NSN is always encouraged. The strategy of *Nayakrishi Andolon* in the maintenance and regeneration of biodiversity and genetic resources is based on some simple rules and obligations between members. These are:

- To remain as an active group member of *Nayakrishi Andolon*, a farming household is obliged to always inform the group of planting decisions and plans for each season.
- Members should harvest seeds collectively, if possible, to make sure that all the valuable seeds of the village have been collected and conserved at the household level and that the quality of the seeds is maintained under the leadership of experienced farmers in the village.
- If seeds of a household are destroyed for any reason, the village leader should be informed and must immediately replenish from other group members or from the CSW (Community Seed Wealth) centre.
- If a farming household is not replanting a variety one year they are obliged as members to give it to a neighbour and make sure that the neighbour replants the variety and both collect seeds for the next season.
- If a farming household does not find anyone to replant a variety, they are obliged to report to the *Nayakrishi* Seed Network and deposit the seeds, if necessary, in the CSW centre.

Seed conservation is an art that largely belongs to women. The leadership of women farmers in seed conservation and community seed wealth centres is almost natural. Women, therefore, are the key elements in building up a national seed network. They are the key actors and leaders in the operation of the NSN. In addition, *Nayakrishi Andolon* has formed a Specialized Women Seed Network (SWSN) comprised solely of women farmers.

The NSN is also linked with *Nayakrishi* Seed Huts (NSHs), households in a village specializing in maintenance of local crop varieties, and Community Seed Wealth Centres acting as community seed banks.

Specialization: Specialized Women Seed Network

To enhance the capacity of the community the Specialized Women Seed Network (SWSN) has

Fig. 19.1. A view of the Community Seed Wealth Centre at Tangail. Maintenance of CSWC demands experience and a thorough understanding of weather and its impact on the regenerative traits of different varieties of seeds. Such knowledge is unique in different ecosystems. The linear structure of language can only partially represent such knowledge. These are areas where accumulated experience and oral tradition of the community are the only guide that has maintained agrobiodiversity for hundreds of years.

been formed. The members of the SWSN are specialized in certain species or certain varieties. Their task is to collect local varieties from different parts of Bangladesh. They also monitor and document the introduction of a variety in a village or locality. They keep up to date information about the variability of species they have been assigned. The responsibility is assigned according to the interest and knowledge of the individuals. The SWSN often shares their findings in large meetings organized by the *Nayakrishi Andolon*. Generally SWSN maintains the following vital information:

- Varieties of a particular crop, their distribution and availability in Bangladesh.
- Community knowledge about the characteristics and traits of a variety, including why a variety has been given certain names by different localities.
- Variety or varieties that have been replaced because of the introduction of a new variety in her area.
- With UBINIG staff, keeps track with the formal research institution about the variety of her interest.

The specialization encourages individuals to be more focused on a few species and as a result they develop valuable knowledge about a particular variety. Since this knowledge is highly valued by the group, she gets immense respect and recognition for her contribution in the process of building up collective spirit and knowledge-sharing. UBINIG staff work with SWSN for oral translation of research reports from published journals coming out of the formal research system. This exercise also effectively contributes to the knowledge base of the women, as they are able to share positive information with other network members.

Village Pool: Nayakrishi Seed Huts (NSH)

Nayakrishi Seed Network is built on the independent initiatives of one or two households belonging to *Nayakrishi Andolon* in the village, who are willing to take responsibility for ensuring that all common species and varieties are replanted, regenerated and conserved by the farmers. These households are known as *Nayakrishi* Seed Huts (NSH). These households in a particular village are linked with the existing NSN structure covering large areas. It often happens that two to three varieties are not replanted in a season in a village. It becomes the responsibility of NSH to replant those varieties. These farmers are specially trained and equipped to handle such a situation. In case the number of unplanted varieties is large during a planting season, she goes to the *Nayakrishi* Community Seed Network (NCSN), a local-level NSN structure. The NCSN ensures that, although a village loses a variety or a species temporarily, the biological and genetic resources are available within the range of the cluster of villages.

Community Seed Wealth Centre

The Community Seed Wealth (CSW) centre is an institution set up in the village that operates as a community genebank for the farmers in the area and articulates the relation between village and the National Genebank. It stores seeds and other planting materials of a wide range of both cultivated and non-cultivated crops found in the villages, using local storage structures (Fig. 19.1). The CSW centres also maintain a well-developed nursery. Apart from seed collecting, storage, preservation, distribution, exchange and regeneration, tasks of the CSW also include documentation and maintenance of overall information on the local plant genetic resources in the area.

The construction of CSW centres is based on two principles: (1) they must be built from locally available construction materials, and (2) their maintenance should mirror the household seed conservation practices. Any difficulty encountered in the CSW centre reflects the problem farmers are facing in their household conservation. The experience of operating CSW centres for the last three years shows that these centres can pay for their costs from the incomes obtained by selling seeds and saplings.

The women members of the NSN run CSW centres. Three women farmers take care of the seed resources of a centre, while two people look after the management of the nursery. The CSW centres receive germplasm from the NSN. All germplasm is registered and relevant information is kept at the CSW centre. To facilitate communication with the National Genebank the accession data are kept in accordance with the standard practice. The farmers who have deposited seeds in a CSW centre can claim the deposited species or varieties any time they want. The members of the *Nayakrishi Andolon* can also collect seed from these centres with the promise that they will deposit double the quantity they received after the harvest. A farming household can decide not to replant a species or a variety in a particular season but may come back after 2-3 years for the same. The seeds are sold to other farmers of the village and the proceeds are used for maintenance of the CSW centres. The CSW centre at Tangail is already self-sustaining. However, the cost of documentation and maintenance of passport data information has been subsidized by UBINIG.

Special care is taken in the processing and storage of seeds in the CSW centres. Farmers' traditional knowledge and practices on seed storage techniques are used. Three factors – container, drying technique and constant monitoring of the weather conditions – are considered to be critical in the safe storage of seeds. Seeds are mainly kept in earthen pots available from the local potters, which are tested by observation for porosity and humidity control. Different types of earthen pots are used for different types of seeds. Seeds that are not going to be planted in the next season require different types of pot with specific features. Coloured glass jars are generally used for vegetable seeds, which is a common practice of the women farmers in the village.

For drying of the seeds, the month of Bhadra (August-September) is crucial and sometimes additional people are required for quick drying of the seed. Management of pest and insect attacks in the container is not a major problem and does not require complicated technology. Usually dried neem (*Azadirachta indica*) leaves are used and containers are sealed with mud and cow-dung. To date the CSW centres have not faced any major problem in the storage of seeds. However, since very little research has been done on the seed conservation under normal household conditions and weather, research on these aspects is built into the management of these centres.

Nayakrishi Natural Resource Auditing Committee

The *Nayakrishi* Natural Resource Auditing Committee (NNRAC) is an institution set up under the NSN that surveys and documents existing species and genetic variability of a village.

The NSN forms NNRACs comprised of the network members for an area within a day's walking distance. Depending on the distribution of the villages, the area under a NNRAC consists of 5 to 7 villages. The NNRAC undertakes an audit of the existing species and varieties of a village, and starts depositing the germplasm in the CSW centres. It also documents resources that must be maintained and managed *in situ*. Species and varieties in the field, cultivated or uncultivated, are documented and registered as the wealth of the community, including their use, as a precaution against unauthorized collecting. Whenever possible, *Nayakrishi Andolon* declares some areas as communally protected and encourages collective action of the villagers to maintain and manage biodiversity in these areas for the benefit of the ecosystem. However, there are no legal regimes available to farmers to ensure such community action.

A large number of species and varieties are not cultivated and thus are more prone to genetic erosion. The conservation and regeneration of biodiversity for these species and varieties are mainly maintained by the overall structure of the *Nayakrishi Andolon*. Every village where *Nayakrishi* is active has a *Gramkormi* (village worker). Apart from networking and campaigning for *Nayakrishi*, *Gramkormis* maintain audits of the natural resources of the village. The information is maintained collectively. This is a vital practice to maintain and manage the local biodiversity.

The *Nayakrishi* farmers can easily be put on alert if any landrace, wild species or variety is noticed as getting eroded or lost.

Linkage with National Genebank

The exploration of the linkage between National Genebanks and the Community Seed Wealth centre has only started, and development of this relationship largely depends on the cooperation of the formal systems. The institutional strength of a farmers' seed network and community germplasm conservation and regeneration can further be demonstrated if the linkage between the formal and informal institutions is formalized. A recent exercise with Bangladesh Jute Research Institute (BJRI) to regenerate jute seeds was very satisfactory. *Nayakrishi Andolon* has received support and collaboration from the scientists active within the structure of the formal systems. The main problem, however, is still bureaucratic hindrance and the research priority of these institutions in a direction that excludes the farmers.

The formal systems have recognized the value of research and activities of UBINIG with the *Nayakrishi* Seed Network. UBINIG has played a very active role in the National Committee on Plant Genetic Resources (NCPGR) and has contributed in their efforts to draft a Biodiversity and Community Knowledge Protection Act as well as a Plant Variety Act. Both of the drafts have been endorsed by NCPGR and accepted by the Ministry of Food and Agriculture for inter-ministerial discussion.

The NSN, through its organizational structure and mediated by the set-up such as CSW, demonstrates profound potential for the conservation and use of local plant genetic resources, not only in collecting and conservation of germplasm. The practice envisages an institutional arrangement and linkages by which a dynamic system of national genetic resources conservation and management can be contemplated. Bangladesh can maintain, manage, regenerate and enhance agricultural biodiversity and genetic resources with very little cost if a farmers' network is linked to the national strategy. The country can not afford to finance a large genetic resources conservation strategy. Moreover, maintenance of expensive genebanks without interactive articulation with the farming community can be dangerous because of the possibility of technological failure and other unseen risks. It also interferes with the evolutionary process of crop development. A strategy based on the dynamic interaction of *in situ* and *ex situ* is the only strategy that is viable for the country, and perhaps for the region.

Conclusions

The activities of *Nayakrishi* Seed Network show enormous possibilities. Nevertheless, a lot of effort and research is still required to consolidate the gains as well as to resolve the gaps and weaknesses. The work in the last few years has created an excellent opportunity to learn with the farmers and to conduct systematic research on the conservation and use of local plant genetic resources. UBINIG, as the source of inspiration of the movement, is focusing on developing the research capabilities of the farmers in this area, both as individuals and as a group. The practice of ecological farming ensures food security and income for the farming communities. *Nayakrishi Andolon* does not end at that level. It is time that we demonstrate clearly that farmers are perfect managers of their seed and genetic resources and are producers of valid and authentic knowledge associated with such resources. It can not be done by romanticization of farming communities, but only through organizational efforts and mobilizations.

20. Local seed systems and PROINPA's genebank: working to improve seed quality of traditional Andean potatoes in Bolivia and Peru

Victor Iriarte, Franz Terrazas Gino Aguirre and Graham Thiele

Introduction

Potatoes are the most important crop of highland farmers in Bolivia, both as a dietary staple and as a source of income. About 80% of the cropped area is sown to landraces. The most widely grown such as Waych'a, Imilla Negra, Imilla Blanca, Wila Imilla, Gendarme, Malkacho and Runa belong to *Solanum tuberosum* subsp. *andigena*. Farmers also grow landraces belonging to the species *S.* × *ajanhuiri, S. stenotomum, S. phureja S.* × *juzepczukii, S.* × *curtilobum* and *S. goniocalyx*. Many of these are of only local importance, but are highly valued because of specific culinary properties. The most floury varieties are boiled in their skins (*Wayk'u* or *Kath'i*) and may be eaten at harvest by family members or when an esteemed guest visits. Some are used in soups and others, such as the Phurejas, are used for special local dishes during religious holidays. Some landraces, in particular of the bitter Luky category, are freeze-dried using artisanal techniques to produce *chuño* and further processed by immersion in a river and sun-drying into *tunta*. Besides their culinary properties many landraces have valued agronomic characteristics, such as the Luky types, which are frost tolerant.

As production has become increasingly market-oriented, a small number of varieties has become more widely grown with a corresponding reduction in the importance of local landraces. Institutional interventions have often encouraged this tendency. In Charazani, La Paz for example, following the severe drought of 1982 when farmers lost most of the seed of the Mayu Rumi cultivar, a development project brought in large quantities of seed of Waych'a and Alpha, a Dutch variety.

Potato tuber seed tends to accumulate viruses and other pathogens over time, a process known as degeneration. As more production goes to market, seed moves longer distances so that pathogens also have become more widely disseminated, aggravating the problem. Farmers reported in a workshop that the seed of local landraces has become increasingly "tired", leading to falling yields and this is a major reason why they grow them less (PROINPA 1998).

Meristem culture and thermotherapy can be used to eliminate pathogens to produce clean seed. Trials comparing the yield of cleaned Waych'a with farmers' seed showed average gains of between 10 and 20%. Yield losses may be considerably higher for severely infected seed. Clean seed is available for Waych'a and a few of the widely grown commercial landraces as well as modern varieties through the formal seed system which covers about 2% of the direct demand for seed. Complex localized flows of seed occur from higher to lower areas. Farmers recognize that seed from higher areas is of superior quality. There, degeneration is slower and seed is likely to be of higher quality. Communities, and even particular farmers, known for the quality of their seed form nodal points in these flows. In Bolivia, farmers in these areas typically acquire a few hundred kilograms of high-quality seed, sometimes from the formal system, often through NGOs or local development projects, to multiply up until they have a surplus for sale (Thiele 1999). Communities specializing in seed production offer a kind of "neighbourhood certification" because purchasers know it has given good results in the past (Scheidegger *et al.* 1989: 9). This cleaned material, moving through the local seed system, eventually replenishes farmers' seed stocks of Waych'a and those few varieties for which formal seed can be obtained. No source of replenishment is available for seed stocks of local landraces, which leads to increasing degeneration and further contributes to their abandonment.

In this article we explain a methodology that PROINPA has developed for returning cleaned seed of local landraces to farmers. We identify the main results achieved, the difficulties faced, make some recommendations for improving the methodology and draw general conclusions about the participatory process.

Methodology

The methodology which was developed had the following steps:
1. Contact with NGOs
2. NGOs prioritize landraces for cleaning with farmers
3. PROINPA prioritizes the requests made by farmers
4. Farmers collect samples of prioritized landraces using positive selection
5. PROINPA cleans up selected landraces and produces prebasic seed
6. Seed is returned to farmers in collaboration with NGOs
7. Production of seed by farmers
8. Wider diffusion of clean seed
9. Monitoring of process.

We will explain the role of PROINPA, the NGOs and farmers in each of these steps. We began work in the Cochabamba Department in 1992 and in La Paz and Oruro Departments in 1993 (see Table 20.1). The method evolved over time as we learned how to do things better.

Contact with NGOs

Initially NGOs in Cochabamba and La Paz that knew PROINPA had facilities for cleaning seed contacted us with requests.

Subsequently, PROINPA contacted other potentially interested NGOs, inviting them to prioritize landraces for cleaning with farmers. We requested that they supply information about the cropped area, uses and agronomic characteristics of the material selected.

NGOs and farmers prioritize landraces for clean-up and provide information about their characteristics

NGOs worked with community-level farmers' organizations known as *sindicatos* to select landraces suitable for clean-up. Each NGO made a request for between 13 and 60 landraces and included the information about the landraces provided to them by farmers.

Farmers' involvement in this stage and the way NGOs facilitated the prioritization varied. In Raqaypampa, Cochabamba the request originated with local experts known as *yanapaj* selected by the *sindicato*. The *yanapaj* approached SENDA, the NGO that worked with them,

Table 20.1. The prioritization, clean-up and multiplication process in La Paz and Oruro

Activity	Year
Information about possibilities of clean-up reaches NGOs in La Paz	1993-94
PIABS, CESA, CPA, contact PROINPA on behalf of *sindicatos* with requests for clean-up	1993-94
PROINPA contacts 11 other NGOs to see if they are interested	1993-94
PROINPA prioritizes on the basis of the requests made through the NGOs	1993-94
Institutions compile farmers' requests	1993-94
PROINPA cleans 15 landraces for La Paz and Oruro	1993-94
PROINPA multiplies seed in greenhouses and protected seedbeds	1994-96
Farmers with support from NGOs and PROINPA multiply 4 landraces in protected seedbeds	1996-97
Farmers multiply 13 landraces in protected seedbeds and 7 in fields	1997-98
Farmers multiply 8 landraces in seedbeds and 8 in fields [†]	1998-99

[†] Data for Oruro are missing.

Table 20.2. Landraces cleaned by PROINPA

Landrace	Species	Department and Province	NGO collaborator	Use of the tubers	Agronomic characteristics	Seed (kg)/ family in 1999[†]	Families with clean seed in 1999[†]
Milagro	S. stenotomum	La Paz Loayza	ACRA	Kath'i	Drought and frost tolerant	150	26
Pali Rojo	S. tuberosum subsp. andigena	La Paz Loayza	ACRA	Kath'i, Chuño	Drought tolerant	150	26
Pali Negra	S. tuberosum subsp. andigena	La Paz Loayza	ACRA	Kath'i, Chuño	Drought tolerant	190	26
Sani Negra	S. tuberosum subsp. andigena	La Paz Loayza	ACRA	Commercial, home use	Drought tolerant	200	26
Bola Luk'i	S. × curtilobum	Cochabamba Tapacarí	ACRUCO	Tunta and Chuño	Frost tolerant	300	46
Chojlla Luk'i	S. × juzepczukii	Cochabamba Tapacarí	ACRUCO	Chuño	Frost tolerant	180	46
Khuchi Sullu	S. stenotomum	Cochabamba Tapacarí	ACRUCO	Wayq'u	Drought and frost tolerant	190	28
Majarillo	S. tuberosum subsp. andigena	Cochabamba Tapacarí	ACRUCO	Commercial, home use	Needs sheltered conditions	240	23
Peraza Luk'i	S. × juzepczukii	Cochabamba Tapacarí	ACRUCO	Chuño	Frost tolerant	180	26
Q'ala Ajahuiri	S. × ajanhuiri	Cochabamba Tapacarí	ACRUCO	Wayq'u	Needs sheltered conditions, drought tolerant	300	25
Q'etu Luk'i	S. × juzepczukii	Cochabamba Tapacarí	ACRUCO	Tunta	Drought tolerant	240	19
Moroq'o Luk'i	S. × juzepczukii	Cochabamba Tapacarí	ACRUCO	Tunta and Chuño	Drought tolerant	180	22
Laqmu	S. tuberosum subsp. andigena	Cochabamba Mizque	CENDA, CEDEAGRO	Commercial, home use	Drought tolerant	45	158
Pucañawi	S. tuberosum subsp. andigena	Cochabamba Mizque	CENDA, CEDEAGRO	Commercial, home use	Virus tolerant	45	158
Isla	S. tuberosum subsp. andigena	La Paz Omasuyos	CESA	Commercial, home use, Kath'i	Drought tolerant	55	46
Wila Imilla	S. tuberosum subsp. andigena	La Paz Omasuyos	CESA	Commercial, home use	Good for storage	65	46
Yacu Imilla	S. tuberosum subsp. andigena	La Paz Loayza	CPA	Commercial, home use	Adapted to low-fertility soils	56	36
Runa Amarga	S. × juzepczukii	Oruro Tomas Barron	COPLA	Tunta, Chuño	Frost tolerant	65	35
Sakampaya	S. tuberosum subsp. andigena	Oruro Tomas Barron	COPLA	Chuño	Drought tolerant	65	35
Mayu Rumy	S. tuberosum subsp. andigena	La Paz Bautista Saavedra	PIABS	Commercial	Virus tolerant and late blight resistant	25	45
Yurak Luk'i	S. curtilobum	La Paz Bautista Saavedra	PIABS	Tunta, Chuño	Frost tolerant	20	45
Luk'i Moroq'o Blanco	S. juzepczukii	Oruro Cercado	UTO	Chuño, Tunta	Frost tolerant	45	35
Pali Morado	S. tuberosum subsp. andigena	Oruro Cercado	UTO	Kath'i and Chuño	Drought tolerant	60	35

[†] 1998 in the case of Cochabamba, data were not collected in 1999.

asking for a solution to the tired seed of Laqmu and Pukañawi cultivars. SENDA contacted PROINPA with the request to clean them. In Charazani, farmers wanted new seed of Mayu Rumi, which they had previously grown, as the cultivars brought in by the development project had subsequently proven very susceptible to late blight.

PROINPA prioritizes the requests made by farmers

In the first year PROINPA agreed to clean all selections. Given the relatively high cost of clean-up (around $US 800 per cultivar), in the second year it was not possible to meet all requests. PROINPA prioritized a smaller number of landraces for each location on the criteria that they were grown over a significant area, and that they covered the full range of uses and agronomic characteristics that farmers required in each area (Table 20.2).

Farmers collect samples of prioritized landraces using positive selection

PROINPA provided a protocol for collecting material, using a procedure known as "positive selection" to reduce virus infection in the potato seed. PROINPA staff made visits at flowering (when plants are marked) or harvest. NGOs coordinated the collecting process. Farmers marked the plants of the selected landraces that had the fewest symptoms of virus infection (such as stunted plants and rolled, mottled or mosaic leaves). At harvest they picked six healthy tubers of each landrace from the plants which had been marked. These tubers were given to PROINPA for cleaning.

PROINPA cleans up selected landraces and produces prebasic seed

Tubers were sown in a greenhouse and checked for viruses using a DAS-ELISA test. Cuttings from the healthiest plants were taken 30 days after emergence. These were treated using thermotherapy and meristem culture to eliminate viruses. Seedlings were grown *in vitro*, and checked using DAS-ELISA to ensure that they were virus-free before producing tuber seed in the greenhouse, or *in vitro* seedlings, which could be planted out in protected seed beds at the Toralapa experimental station (Iriarte *et al.* 1998).

Protected seedbeds played a central role in the multiplication process. The seedbeds are boxes made of adobe bricks or other local materials about 10×1.5 m in size. They have a layer of clean soil with organic fertilizer at the bottom. The seedbeds are carefully watered and tended. They can be covered with plastic sheeting or other materials when there is a risk of frost or hail. When using small tubers (10-15 g), 3-5 kg of seed are usually required for a rustic bed of around 17 m^2, which normally produces from 40 to 60 kg of high-quality seed to be multiplied the next year in the field. This is about double the multiplication rate that could be obtained under normal field conditions (Aguirre *et al.* 1999).

We organized a workshop in Toralapa for farmers and extensionists from the NGOs who were involved in the project (PROINPA 1998). They visited the laboratory and other installations used for cleaning seed. We explained the process using analogies with human diseases. Farmers commented that the old and tired seed had become "a child again" (*wawachaska*).

Seed returned to farmers in collaboration with NGOs

A meeting was arranged with the farmers' *sindicato* (in coordination with the NGO) to return the cleaned landraces. At the meeting PROINPA handed the seed over to the NGO who gave it to the leaders of the *sindicato*. All three parties signed a short contract to confirm the handover. The leaders explained to the rest of the *sindicato*'s members that this was the seed that had been collected earlier and was now being returned. Because the multiplication process in potatoes is slow, typically only a few kilograms of seed of each landrace were handed over. Given the small quantities involved and the risk of loss, PROINPA agreed to supply similar quantities of seed for multiplication in the community for two subsequent seasons. Initially, PROINPA covered seed costs with project funds, subsequently the NGO paid, and now farmers' organizations are paying for the cost of this seed.

Fig. 20.1. Andean farmers comparing improved seed stocks of their native varieties.

Production of seed by farmers

As the first clean seed was being returned, farmers in Japo, Cochabamba asked how they could protect this valuable seed. PROINPA suggested the protected seedbeds already being used in the experimental station as an option. Subsequently, in coordination with the NGOs, 282 farmers were trained in the construction and management of the beds and all of the clean seed received by the *sindicatos* was first multiplied in the seedbeds before being planted out in carefully managed fields for seed production. Farmers were also trained to use sprouts from seed potatoes, which can also be planted, in seedbeds to increase the rate of multiplication. Extensionists from the NGOs provided technical support during the multiplication process with guidance from PROINPA.

Wider diffusion of clean seed

The distribution of the first generations of seed produced was agreed with the *sindicato* and in some cases with the NGO. Subsequently there was a wider informal distribution of seed by sale or gift to relatives and neighbours.

Monitoring of process

PROINPA kept careful records of the amount of seed supplied, and production in seedbeds and open field. Farmers were curious to know how well this seed performed. They conducted their own trials comparing the yield of seed originating from the cleaned material with their own seed of the same variety in the same field. PROINPA monitored these trials and analyzed the data.

Results

In all, 23 landraces were cleaned up and returned to farmers in 14 communities in the Departments of La Paz, Oruro and Cochabamba. Yields in the seedbeds ranged between 0.3 and 4 kg/m^2, with a mean of around 2 kg/m^2 for *in vitro* seedlings, and from 2 to 9 kg/m^2, with a mean of around 3 kg/m^2, for tuber seed. This is more than double the normal field yields. Farmers were very pleased with the protected seedbeds.

Farmers' trials in Tapacari and Araca showed that, in general, clean seed yielded 26-40% more than their own seed (Fig. 20.1).

As a result of using protected seed beds and field multiplication, more than 100 tonnes of clean seed is in farmers' hands and is being multiplied up for wide-scale planting.

Farmers are convinced of the benefits of cleaning seed and are requesting clean-up of other landraces. In Araca (La Paz) farmers have requested it for Chejchi Imilla, in Eucalyptus (Oruro) for Polonia, and in Mizque (Cochabamba) for Runa Criollo.

Difficulties

Because of the large number of landraces and differences in local names correct identification is often difficult. In the first year of clean-up PROINPA ran electrophoretic tests on Laqmu and found it to be identical to Sani Runa which had been cleaned previously. It was decided to use the clean Sani Runa rather than clean the seed farmers had provided. When farmers were shown the material and told it was Laqmu they pointed out slight differences in tuber skin colour and foliage. In fact, farmers were right and Laqmu was not the same as Sani Runa. As a result it was decided to always use the material farmers supplied rather than trying to use similar material already in the germplasm collection.

Clean-up took a lot longer than anticipated. Some of the samples that in farmers' fields had shown symptoms of viral infection were virus-free according to the DAS-ELISA tests conducted before thermotherapy. Apparently there were Andean variants of the viruses which these DAS-ELISA kits could not detect. The clean-up and multiplication programme was put on hold until this problem could be resolved. Some farmers became frustrated with the 2-3 year gap between priorization and return of clean materials. The clean-up process has now been streamlined and the period between supply of samples and the return of clean material reduced to 14 months.

In some instances cleaned landraces differed considerably from farmers' seed. Wili Imilla when infected with virus has completely red tubers but has white patches when clean. Farmers did not believe that it was actually the same cultivar when it was returned to them. In one of the farmer's trials, with Chojilla Luk'I, clean seed produced many more tubers per plant than farmers' seed but the time to physiological maturity was lengthened in virus-free seed. A frost fell towards the end of the season and because the plants grown from the clean seed were less developed they yielded substantially less in kilograms than the farmer's seed. The farmer, however, was still happy with the outcome because he had a much larger number of seeds to plant the next season.

In one instance, Chajlla Luki, clean-up did not seem to improve yields at all. This shows the value of farmers' trials in measuring yield gains.

Finally, in Raqaypampa which was the lowest altitude site at which we worked (2800 m asl), the seed beds failed because under warmer conditions they encouraged late blight infection. Here intensively managed open seedbeds would be preferable. PROINPA no longer promotes seedbeds below 3000 m asl.

Recommendations

The process of prioritizing local landraces and gathering information about them with farmers could be improved using community-level workshops and techniques of participatory rural appraisal. For example, the *sindicatos* could map the distribution of landraces in their agro-ecologies, draw tubers with descriptions of their characteristics and use ranking methods or pairwise comparisons to select the most appropriate ones for cleaning.

In general, *sindicato* meetings are made up of male household heads. Working through *sindicato*s led to a male bias in planning seed multiplication. Throughout the Andes, women play a central role in seed management (Tapia and de la Torre 1993). In the communities where we worked women were responsible for obtaining good-quality seed from relatives, in a neighbouring community or in the local market. Women have a vested interest in the success of the seedbeds because they will save time that would have been spent searching for quality seed. Subsequently, the gender issue was addressed, as we saw above. In the future, training in seedbed construction and management, and decision-making about seed multiplication, should explicitly involve both men and women as appropriate.

During the workshop that was held in 1997 farmers recognized in PROINPA's germplasm collection some landraces which they had previously grown and which had been "lost" (PROINPA 1998). A next step could be to promote access to this material and clean it up for return.

Conclusion

The strategy underlying this project is to build on the strengths of the local seed system by supplying small quantities of clean seed to farmers, who multiply it up and let "neighbourhood certification" work to distribute this seed more widely. Since seed degenerates quite slowly, small injections of clean seed should be sufficient to improve seed quality for large numbers of farmers. Evidence so far is that subsequent generations of seed supplied by PROINPA are now beginning to find their way into the hands, and fields, of large numbers of farmers through this local system and that in the case of the PESEM it is also improving seed quality in formal seed being supplied to institutions. To evaluate the full impact of the project will need close monitoring of the diffusion of clean seed through the informal seed system.

The project has helped maintain biodiversity and has raised productivity in farmers' potato fields and created an awareness of the importance of good-quality seed. Because potatoes are such an important feature of farmers' lives it has also improved living standards in a more qualitative sense. Señora Filomena Chanvi summed it up nicely: "my mother used to prepare the lunch we'd eat after working in the fields, she would spread her *aguayo* (large fabric used for carrying) on the ground and put a white cloth on it where she would spread the *tuntas*, *chuños*, the P'iñus cooked in the oven and the other potatoes the Imillas and on top of that she'd put a delicious roast prepared with hot stones. We'd eat those P'iñus first, then the roast meat. Those good times are being lost together with our native varieties, which are disappearing from this place. This work which we have done with our strength and in collaboration with you has helped us to produce more than 5 tonnes of seed and this process will help us to bring back the varieties which we are losing."

The combination of a research and development institution (PROINPA), NGOs and farmers' institutions has been fundamental in this success. Each has made a specific contribution. PROINPA has provided technologies (seedbeds and clean-up process) as well as a methodology for returning clean seed to farmers, the NGOs have facilitated the process locally and provided a bridge between farmers' organizations and PROINPA and the *sindicato* have managed the process at the community level.

Acknowledgements

The advances reported here would not have been possible without the collaboration of ACRA (Asociación de Cooperación en Africa y America Latina), AGRUCO (Agroecología Universidad Cochabamba), CENDA (Centro de Comunicación y Desarrollo Andino), CESA (Centro de Servicios Agropecuarios), CNS (Consejo Nacional de Semillas), COPLA (Centro Orureño de Planificación), CPA (Central de Productores Agropecuarios), Fundación Contra el Hambre, PIABS (Proyecto de Investigación Agraria Bautista Saavedra), PROSUKO (Programa Interinstitucional Sukakollo), SEMTA (Servicios Múltiples de Tecnologías Apropiados), Wiñaymarka, UTO (Universidad Técnica de Oruro), Visión Mundial and the farmers who work with them.

> ### Ayjadera and Japo: formal and informal seed multiplication
>
> Farmers choose to organize multiplication in very different ways. We look at two cases here.
> - The *sindicato* in Ayjadera, in the Araca Province of La Paz, works with the NGO ACRA. The *sindicato* delegated responsibility for seed multiplication to the PESEM - a small seed-producing enterprise that had been established in the community – as it was already experienced in seed production. PROINPA supplied 250 *in vitro* seedlings for each of Sani Negra, Pali Rojo, Pali Negra and Milagro in 1996-97. The 24 members of the PESEM planted out these seedlings in the seedbeds. In 1998-99 they harvested about 2 tonnes of each variety. They have certified this seed in the formal seed system. They are selling this certified seed in the local market and to NGOs and farmer organizations in other areas. In this case the agreement with the *sindicato* was that they should offer seed to other members of the *sindicato* on a sale basis.
>   In Ayjadera, the PESEM is made up of men. In 1996-97, only men took part in training activities with the seedbeds: the construction of the seedbeds is heavy work where they actively participated. PESEM members said that looking after the beds was light work, and progressively gave more responsibility to women for this task. Women took part in the training the next year. Both men and women are actively involved in selling seed.
>
> - In Japo, Cochabamba the *sindicato* was given about 3 kg of seed of each of Majarillo, Moroqo Luk'i and Qetu Luk'i in 1995-96. The *sindicato* picked Don Dionysio Chambi, a well-known seed producer in the informal system, to multiply this material. Dionysio manages around 15 varieties in seven different fields for seed production in the collective rotations (*aynokas*) which are still prevalent in Japo. Dionysio planted the seed in four protected seedbeds. His yields were well above those obtained in the experimental station (125 kg of Moroq'o Luk'i from one seed bed, the equivalent of 90 t/ha). Dionysio gave half of the seed back to the *sindicato*, as had been agreed, and this was shared amongst 46 farmers. By 1997-98 Dionysio alone had over 1 tonne of seed of the three cultivars.

References

Aguirre, G., J. Calderon, D. Buitrago, V. Iriarte, J. Ramos, J. Blajos, G. Thiele and A. Devaux. 1999. Rustic Seedbeds, a Potential Bridge Between Formal and Traditional Potato Seed Systems in Bolivia. CIP Program Report 1996-98. CIP, Lima, Peru.

Iriarte, V., A. Badani, C. Villarroel, G. Aguirre and E. Fernández-Northcote. 1998. Priorización, limpieza viral, producción de semilla de calidad básica y devolución de cultivares nativos libres de virus. PROINPA, Cochabamba.

PROINPA. 1998. Memoria. Primer Encuentro Taller sobre el Mantenimiento de la Diversidad de Tuberculos Andinos en su Zonas de Origen. PROINPA, Cochabamba.

Scheidegger, U., G. Prain, F. Ezeta and C. Vittorelli. 1989. Linking formal R&D to indigenous systems: a user oriented seed programme for Peru. ODI, London.

Tapia, M.E. and A. de la Torre. 1993. La mujer campesina y las semillas andinas. FAO, Lima, Peru.

Thiele, G. 1999. Informal potato seed systems in the Andes: Why are they important and what should we do with them? World Development 27(1):83-99.

Section V

Increasing public and policy awareness of conservation and use of plant genetic resources

21. Overview

Bhuwon Sthapit and Esbern Friis-Hansen

Within the recent past the concept of agricultural biodiversity has passed from the domain of the academician to the widespread attention of the politician and popular press. The key international institutions concerned with biodiversity conservation have underlined the important role of public awareness and education in fulfilling their objectives. They also promote the development of international collaboration in education and awareness-raising. The Convention on Biological Diversity obligates contracting parties to "cooperate, as appropriate, with other States and international organizations in developing educational and public awareness programmes" with respect to the conservation and sustainable use of biological diversity.

Public education and awareness (PA) programmes encourage strong and sustained support for local and national plant genetic resources management activities by promoting the role that genetic resources play in development. Public awareness is needed at different levels, from farmers to policy-makers. Public awareness campaigns waged by either NGOs or governments can shape public opinion. The key element is cultivating interest among trendsetters. In all societies, opinion leaders expose and popularize new issues, as well as catalyze practical action to address them. These opinion leaders – village elders, radio announcers, television commentators, actors/actresses, newspaper editors, famous poets, popular entertainers, athletes, religious leaders and popular local politicians – can make biodiversity messages more compelling to the general media. For this reason, a great majority of public awareness-raising activities have targeted policy-makers in developing countries. Secondary audiences include media and those non-governmental organizations that are in a position to influence government policies with regard to genetic resources conservation and management at the national, regional and international levels. Nevertheless, the most important group of target audiences have been ignored so far because of the lack of appropriate participatory methods. They are farming communities and consumers from economic or agro-ecologically marginal areas.

Such farming systems are important sources of diversity for the majority of farming communities and today most genetic diversity rich areas are found in geographically isolated regions. A majority of rural farming communities do not yet have access to popular culture such as television, video and reading materials. In such areas, rural farmers, particularly women farmers, often cannot read and write. These people do not have access to new information and knowledge on the changing scenario of PGR. When opportunity arises, they even have difficulty understanding the jargons of most local professionals, though they share the same common language. There is a lack of participatory public awareness tools by which the difficult concepts of conservation can be explained in informal ways, so that these custodians are further sensitized and motivated to adopt sustainable agricultural development practices.

In recognizing the necessity of reconciling conservation priorities with the socioeconomic needs of local communities, farmers and communities will need to be involved in on-farm conservation as much as possible. Since most conservation projects are largely community based and quality participation of the community is essential to understand the scientific basis of *in situ* conservation, there is a need for strong public awareness activities at the grassroots level as well.

Initially formal meetings, orientation training, information dissemination through fliers in vernacular language and personal contacts with key people are the conventional methods used. Despite that, it has been realized that our messages are not reaching important members of farming communities, particularly women farmers.

The case studies in this section illustrate various innovative participatory approaches to reach rural communities and to strengthen capacity of grassroots institutions. Tools such as village workshops, social resource mapping, diversity fairs, rural poetry journeys, folk song competitions, roadside dramas, food fairs, etc. have already been tested, refined and revalidated at different sites. While the avenues for strengthening public awareness activities vary with locations and culture, every community has numerous communication tools at its disposal.

Community participation is a central issue in Chapters 22, 23 and 24, which discuss experiences in Nepal, India and Bangladesh with on-farm conservation. It is increasingly being realized that community participation can be strengthened by several means: by sensitizing the farming community and consumers, by developing markets for local products or by providing market incentives, by improving farmers' varieties, and by adding benefits through policy incentives. The innovative approaches used in Nepal and India could be scaled up in other countries in different sociocultural contexts. The rationale behind this is that the strategy for *in situ* crop conservation will only succeed if indigenous communities and their grassroots organizations are involved at different stages, and their needs and problems are shared and addressed.

Chapter 23 illustrates how ethnobotanical research can generate increased awareness among local communities about the value of their plant genetic resources. The research was carried out in Northern Nagaland in India and had a highly participatory methodology, including: a thorough training of local youth as research facilitators; stakeholder analysis and involvement of key informants, including elders, women, village administrators, immigrants and children, and conducting community documentation workshops.

Furthermore, the general public and policy-makers are increasingly aware of the scope and seriousness of the disappearance of the earth's genetic heritage. Piecemeal efforts at genetic resources conservation that do not feature a long-term commitment will accomplish little. The choice of appropriate PGR management strategies requires clear policy decisions about the appropriate mix of formal and informal sector contributions to local crop development and management. This will require good policy for PGR management and conservation, but good policy always depends on good information which can be generated through innovative approaches to policy research. However, participatory policy research is a new and evolving field. Some initial policy work from Nepal is discussed in Chapter 25 for further development.

22. Adding value to landraces: community-based approaches for *in situ* conservation of plant genetic resources in Nepal

Dipak Rijal, Ram Rana, Anil Subedi and Bhuwon Sthapit

Introduction
Raising awareness is the first step in promoting conservation and use of local plant genetic resources. It adds value to the local crops and encourages consumers (both rural and urban), development workers, policy-makers and farming communities to conserve and make continued use of these crops. Various methods such as diversity fair, rural poetry journeys, street dramas, folk song competition, conservation essay competition, etc. have been employed to create awareness at different levels. The concept, process and experiences gained out of locally employed tools in creating awareness are discussed in this paper.

Diversity fair
Diversity fair is an effective medium to sensitize communities on the value and use of conserving local landraces, and to assess genetic diversity at community level in Nepal. It also has been used in other countries in the region, such as India and Bangladesh, and is a common approach in Latin American countries (Tapia and Rosa 1993). Unlike conventional agriculture fairs, diversity fair focuses on local food crops and vegetables, and their landraces. The associated knowledge base of the farming community is also documented in the process. The diversity fair provides a good forum and opportunity to individual farmers and farming communities to display their crop genetic wealth and indigenous knowledge held through generations, and to get recognition for these valuable resources through diversity awards. The use of diversity fair is, therefore, manyfold (Box 22.1). The diversity fair discussed here is based on the experiences in organizing such fairs in three *in situ* project sites in three distinct agro-ecological zones of Nepal.

Process/steps involved in the diversity fair
Diversity fairs are organized with the full participation of the farming community, which takes a lead role throughout the process, while researchers/development workers take a facilitating role and provide technical support. Steps undertaken in organizing diversity fairs are discussed below.

Box 22.1. Uses of Diversity Fair

- Locate genetic diversity and custodians
- Identify value and status of landraces
- Understand reasons for growing landraces on-farm
- Strengthen community awareness on their resources and their control of them
- Enhance process that transfers indigenous knowledge to a new generation
- Facilitate sharing of knowledge and genetic resources among a wide range of PGR users
- Raise awareness of stakeholders of PGR from grassroots through to policy levels
- Fine-tune linkages between *in situ* and *ex situ* conservation and use.

Conceptualizing diversity fair with the farming community. The idea of organizing a diversity fair is first discussed with the farming community by organizing a village workshop. The participating farmers are briefed on the values and uses of diversity fair, and discuss whether or not they would like to hold such a fair in their community. Researchers, development workers and community-based organizations (CBO) facilitate this process.

Organizing committee and defining roles and responsibilities of various actors. An organizing committee comprised of representative farmers of the community, researchers and development workers is formed, and roles and responsibilities are assigned to the committee members for various tasks. Community members and the CBO take responsibility for providing space to stage the diversity fair, inform participating farmers and provide other logistic support. Researchers and development workers assume responsibility for publicizing it to a wider community and facilitate the development of a technical framework for presentation, monitoring and evaluation of display of genetic materials in the fair.

Developing working procedures for diversity fair. A date for diversity fair is first fixed in consultation with representatives of the farming community. The organizing committee then agrees on:
- procedures for participation in the diversity fair, signing up, regulations, etc.
- crops and other species and their varieties to be included in the diversity fair
- farmers' groups who would display their seeds, planting materials, etc. in the fair
- evaluation criteria and process for the displayed material
- awards – items and amount.

The sequences of activities required to stage a diversity fair is carefully worked out and responsibilities are assigned to the organizing committee members.

Publicizing the event. The diversity fair is publicized using local media (newspaper and radio) and through farmers' network in the community. Information about the fair is relayed to farmers, researchers, development workers, traders, politicians and local government representatives, who are invited to participate in the fair.

Staging diversity fair. The organizing committee, with the active support of CBOs and farmers' groups, stages the diversity fair on the announced date and at a location suitable to the farming community. Stalls for display of genetic materials are constructed in advance and allocated to the participating farmers' groups.

Participatory evaluation of the display entries, and awarding winners. An evaluation team, consisting of knowledgeable farmers, researchers and development workers, assesses the stalls, scores the better entries and declares the winning farmer(s) and/or farmers' group(s) for awards. In the diversity fair organized by the IPGRI *in situ* project, the following evaluation criteria were used:
- number of varieties/landraces of targeted crop species
- quality and reliability of knowledge base of the group, on the materials displayed
- attractiveness of presentation
- ways of displaying the materials
- involvement of community members of different age, gender and status groups in the process.

In situ *characterization of landraces entries displayed in the diversity fair.* The diversity fair provides a large collection of landraces and gives a good idea of crop diversity existing in a particular community. The collected landraces, however, have to be checked for their authenticity as distinct varieties through on-farm characterization. First, all landraces registered by a group are listed separately along with the names of farmers who provided seeds for the diversity fair. In each cropping season, an experienced farmer from each group is selected to locate land parcels with individual landraces listed. Farmers who had provided seed samples for the diversity fair are taken as reference farmers. Agromorphological characteristics for each listed landraces are recorded and seed samples required for assessing post-harvest traits are collected.

Effectiveness of the biodiversity fair
The diversity fairs organized in three sites of the *in situ* project have been very useful in raising awareness on the value and uses of local crop landraces within as well as outside the community. Appreciation of farmers' crop varieties through display and awards has encouraged the farming community to realize the importance of conserving and maintaining greater diversity on-farm. The seeds collected during the fair are now kept together as a seed museum in the community for display and learning for others. The fair provided a good opportunity for farmers, researchers and development workers to share information and planting materials of local crops. The seeds of landraces have been given to the national genebank. For the project, diversity fair has acted as an entry point to reach the farming community and get farmers' participation in the programme.

Community Biodiversity Register (CBR)
CBR is a register maintained by a local community or institutions to record the existence of biodiversity and associated knowledge base of communities at local level. This is a simple database that includes documentation of (1) name of farmers' variety, (2) name of farmers who hold seed, and (3) reasons for cultivation. In Nepal, CBR is being evaluated as a participatory monitoring tool of local crop diversity and provides feedback to formal and informal sectors about the status of local biodiversity. CBR provides an opportunity to develop options for conservation and use and several potential uses are identified. The idea arose when the names of local diversity and their locations were recorded during diversity fair. Later, this database was conceptualized as a registration of community biodiversity.

Currently, the *in situ* Project of Nepal is working with local communities to develop a local system of CBR that meets the objectives of farmers and national and international institutions on PGR conservation and use. CBR could be used to monitor the status of genetic erosion of agrobiodiversity at a local scale but this activity should benefit local farmers in exchange and production of seed. In a true sense, CBR is a decentralized way of managing community genebanks where the community participates, each member providing information on the germplasm and knowledge they hold. It is essential to develop ownership of the activity so that local communities initiate CBR for local purposes to list all local existence of and uses of biodiversity found in the village or community.

Gramin Kabita Yatra *(Rural poetry journey)*
The *Gramin Kabita Yatra*, literally meaning rural poetry journey, has been used as a new public awareness tool to convey value and uses of local plant genetic resources among rural and urban populations. A combined team of both national and local poets and poetesses travels together in the rural areas interacting with the farmers, observing on-farm biodiversity of landraces and translating their feelings and experiences into beautiful poems and songs. This has been a unique event where poets and poetesses have been mobilized to document the value of plant genetic resources in songs and poems. The songs and poems written along the journey are recited to the community members to convey the message of conservation and utilization of local landraces.

Processes/steps involved in rural poetry journey
Reputed poets living in the region and the local poets from villages along the route of the rural poetry journey are identified and met by the project staff. A meeting of these poets and poetesses is called to discuss the organization of the journey. Specifically, the following steps are undertaken (adapted from Rijal et al. 1999):
- define process and identify the team of poets and poetesses including local participants
- orientate poets and poetesses on objectives of the mission
- decide on dates and venues, inform communities and hold press conference
- make site visits and hold interactions to understand local dialects and values
- write poems and songs using colloquial terms
- recite poems in front of community members gathered locally
- recite poems at wider gatherings in the region
- publish them in newspapers and in the form of a published collection for wider dissemination.

The project has published the collection of poems and songs written during the journey. The collection has been named *Sampada*, meaning wealth, and the publications have been given back to the farming community. The community can sell these publications to raise awareness and raise funds for the conservation of local landraces.

Effectiveness
It is rather difficult to assess whether the tool has been effective or not. However, indicators such as (1) effective participation, (2) repeated requests for periodic journeys, (3) ability to document village identities along with the value of local resources, and (4) appreciation of the tool as innovative confirm the effectiveness of this tool in raising public awareness on the importance of local landraces.

Gramin Sadak Natak *(Rural Street Drama)*
Gramin Sadak Natak, literally meaning Rural Street Drama, is one of the locally popular innovative media commonly used to demonstrate social and local systems by a group of actors and actresses. Live shows with sensible topics are often accepted and found effective.

It is not only used for entertaining people but also to address social ills and deviations. These days drama is increasingly being used for creating awareness as well as to improve communication skills more effectively. To make drama more effective, locally popular dialects and terms are often chosen.

Processes/steps in organizing street drama
The following steps have been used in organizing street drama:
- Identification of local writer and drama group
- Orientation to drama group on objectives of the project and messages to deliver
- Conceptualize the theme of drama with the drama group
- Write-up, review, rehearsal and selection of location for staging drama
- Informing communities and hold press conference to publicize the event
- Preparation and staging drama in the rural settings
- Follow-up for feedback and further improvement of the content and presentation of drama

Effectiveness
The first drama staged in one of the *in situ* project sites was a tremendous success in drawing attention of the community members to conservation issues (Fig. 22.1). The real-life expression of the actors/actresses using local dialects, words and mimic actions motivated farmers and

other audiences to understand the value and use of local landraces on the verge of extinction. Farmers and CBOs appreciated both content as well as the actions of the actors. The street drama has thus been proved effective in creating awareness at the grassroots level. The effectiveness, however, depends upon the content and the way it is presented and also whether or not it reflects the real-life situation of the village.

Folk song competition on plant genetic resourcesconservation issues

In Nepal folk songs are a powerful means of expressing ones' feelings and views among the rural communities. Rural people enjoy folk songs more when they are based on local dialects and when they reflect the real-life situations. In the rural communities, men and women sing quite often either at collective works or during special ceremonies. They, especially women, are good at composing songs that fit well into popular rhythms. There are also some special ethnic groups, like Gandarva or Gaine. These people, by profession, are expert in composing and reciting folk songs *in situ*. These folk songs can be a very useful tool to deliver conservation messages to farmers as well as to communities at large.

Processes/steps in folk song competition
Organizing a folk song competition involves the following steps:
- Analyze strength of folksong as a tool in delivering conservation message
- Identify individuals and/or farmers' group
- Declare awards and work out evaluation system
- Identify occasions, dates and venue (location)
- Manage musical instruments and musicians
- Inform farming communities about the event and arrange for their participation
- Stage the competition and give away awards in a public ceremony.

Effectiveness
The competitive folk song has motivated local musicians and writers to compose songs related to genetic resources conservation and use. Naturally, some of the groups are not very happy when they are unable to win the prize but it encourages them to perform better in the future. Follow-up is required to get regular feedback necessary to make such programmes more effective.

Essay competition on plant genetic resourcesconservation issues

Essay writing in schools and colleges on a given topic is quite common. Writing essays not only improves one's writing skills and logical thinking but also encourages authors to review relevant information and literature, discuss with experts, and make minute observations on the real-life situation. Essays on various aspects of biodiversity, therefore, encourage young generations to learn and understand more about such issues and can help create awareness in their circle.

Processes/steps in organizing an essay competition
The steps involved in organizing an essay competition on conservation issues are:
- Identify objectives, analyze the strength of the tool with target groups
- Discuss the norms, areas, topics and schools or individuals to be involved
- Formulate evaluation criteria and form a team
- Decide on awards and prepare certificates
- Invite essays from the school on the given topic related to biodiversity
- Assess essays
- Give awards to winners.

Fig. 22.1. Gaun Ko Katha Yastai Huncha Hai –" Such is the happenings of a village", rural drama, Nepal.

Effectiveness

The approach has helped create awareness on the importance of PGR conservation among the new generations of the community, who obviously are the future conservators. Among the participating schools, conservation of local crop varieties has become a matter of discussion among the students and teachers alike. They feel proud not only of winning prizes but also of being recognized for their better understanding of issues in PGR conservation. In addition, it has helped to create awareness in the community at large.

Adding benefits through market incentives

Steps for adding benefits to local landraces through market incentives

Farmers maintain landraces because of the various benefits – market and non-market – which they obtain through their uses. Other stakeholders, traders, processors and consumers also assess values of landraces for their own purposes. For conservation to work, we may have to increase these values. Some of the ways to do this are presented below.
- Identify niche market in the local community, e.g. local shops, restaurants, hotels, etc.
- Understand the incentive structure for production and marketing, e.g. in terms of quality, cultural value, market pricing, etc.
- Identify potential landrace-based products, through the acquisition of indigenous food-processing knowledge, to suit the requirements of various consumers.

Promoting means of market promotion

The following means of market promotion for local crop produces have been initiated:
- Increase consumers' awareness on the value of local products by providing certificate of origin and putting variety's local name as trade name
- Generate, document and disseminate information on food culture (e.g. cookbook)
- Analyze the nutrient content of food products made of landraces
- Develop local cuisine using landrace products and improve packaging
- Link market with ecotourism
- Link products of landraces with organic farming and health products
- Include local food on menus in restaurants, hotels, student messes and other food outlets.

Effectiveness

The initial results of market promotion activities pursued in the *in situ* project in Nepal have shown that adding value through market incentives has a positive impact on the on-farm conservation of landraces. The market outlets for food products made of landraces are gradually increasing and more and more local entrepreneurs are emerging in the business of such products. Consumers' habits in buying such foods are also increasing. The demand for one scented rice grown in the Pokhara valley is so high that it is mostly sold directly to consumers from the farm at quite a high price. As a result, the area under this variety is gradually increasing in many farming communities. A similar effect can occur with other landraces provided they meet taste and other requirements of the consumers.

Conclusion

Local crop varieties can have value specific to a location or to an ethnic group. Rice landraces like Anadi and Sathi, for example, are maintained because of their cultural and medicinal values. On the other hand, genetic resources may erode when they are not competitive to the options a farmer or a community might have. Creating incentive through different means – cultural, economic and ecological – of adding value to landraces is, therefore, important in on-farm conservation of these crops. Raising awareness among consumers and producers, and creating market incentives for landraces are effective strategies in *in situ* conservation of local plant genetic resources.

References

Rijal, D.K., B.R. Sthapit, R.B. Rana, P.R. Tiwari, P. Chaudhary, Y.R. Pandey, C.L. Poudel and A. Subedi. 1999. Options for adding values and benefits of local crop diversity through a non-breeding approach: Experiences of Jumla, Kaski and Bara sites in Nepal. Paper presented at the Third Participants Meeting of the Project, "Strengthening the Scientific Basis for *in situ* conservation of Agriculture Biodiversity On-farm", co-hosted by NARC, LI-BIRD and IPGRI held at Pokhara, Nepal, 5-12 July 1999.

Tapia, M.E. and A. Rosa. 1993. Seed fairs in the Andes: a strategy for local conservation of plant genetic resources. Pp. 111-118 *in* Cultivating Knowledge: Genetic Diversity, Farmer Participation and Crop Research (Walter de Boef, K Amanor, K Wellard and A Bebbington, eds.). IT Publications, UK.

23. Participatory ethnobotanical research for biodiversity conservation: experiences from Northern Nagaland, India

Archana Godbole

Introduction

The participatory approach presented and discussed in this chapter has been designed in connection with a 3-year ethnobiological study of plant genetic resources conservation and use strategies among Konyak Nagas from Northern Nagaland in India. The study was undertaken in four villages in the lower and upper Konyak areas. The approach has helped to establish linkages between various stakeholders and has initiated a process of conserving and further developing indigenous knowledge and local plant genetic resources.

The research project has three objectives:
- to analyze the plant-people relationship through ethnobiological studies
- to assess the significance of current indigenous knowledge, resource-use pattern and strategies of plant genetic resources conservation
- to specifically study the home gardens and agro-ecosystem management of *jhum* agriculture.

Nagaland is one of the seven North Eastern states of India, showing a wide range of cultural as well as biological diversity. Konyak Nagas occupy the northern hilly district of Mon in Nagaland. Konyak are the least economically prosperous among the Naga tribes and yet they are tremendously prosperous in biodiversity owing to their elaborate conservation strategies which have been successfully developed over generations. The high diversity of plant genetic resources has been maintained and managed by Konyak Nagas through well-developed home gardens and *jhum* practices. Konyak home gardens, locally known as *pesha*, exhibit high diversity in size, shape, location, inter- and intra-specific diversity among species cultivated, various types of management, organization and utilization patterns (Table 23.1).

Documentation of PGR from different natural resource management systems has been carried out in nine villages (listed in Table 23.1) of which four are from upper Konyak and five are from the lower Konyak region. The informal participatory approaches helped to collect this comprehensive data within a short time span of 6-9 months. A few components of PGR (such as number of species in the home gardens and the varieties of major crops in *jhum*) from these villages have been compared in Table 23.1.

Although the traditional system of home gardens is common in all the Konyak villages, it has been perceived quite differently in different villages. The area, selection of plants to be grown, architecture, management, etc. change significantly according to the need and taste of the owner as well as the village. The home gardens from the villages of Changlangshu, S'Tangtin and Tanhai show the maximum number of species cultivated, while villages such as Liangnyu and Wakching have poorly maintained home gardens with low species diversity. The latter show a low total number of species cultivated in spite of the area available per household.

Data collected on major crop varieties of *jhum* cultivation clearly show that upper Konyak villages have more varieties of millet and maize than lower Konyak villages. Maize is the main crop in most of the upper Konyak villages, while lower Konyak villages cultivate rice as their main crop. The upper Konyak villages cultivate more rice varieties than the lower Konyak villages.

The effects of industrialization and the Green Revolution in India are visible even in the remote corners like Nagaland where, with the introduction of the wet rice cultivation methods, new hybrid rice varieties were brought in. Selection of these high-yielding varieties led villagers

Table 23.1. Useful crop varieties in Konyak Nanga home gardens, Mon District, Nagaland, India

Village name	Number of species in home gardens	Number of varieties		
		Rice	Maize	Millet
Upper Konyak				
Changlangshu	112	19 (+3)†	6	14
Münnyakshu	97	9 (+2)†	5	14
Jakphang	72	14	7	6
Longchang	69	30	13	16
Lower Konyak				
Tanhai	122	15	5	6
S'Tangtin	105	17	4	8
Mon	87	14 (+1)†	5	6
Wakching	76	6	4	4
Liangnyu	45	13	5	4

† Figures in parentheses indicate the number of lost varieties.
Source: Participatory documentation workshops carried out in nine villages in 1998.
All data are based on farmers' perceptions and have not been physically verified.

to abandon a few local crop varieties of *jhum* that require laborious cultivation methods. Consequently, a few of these have been lost over time.

The data collected from both upper and lower Konyak areas helped in designing the future participatory approaches for PGR conservation. Data analysis clearly indicated the need for more concrete interactions within the Konyak communities of both regions to understand the loss of varieties and their methods of cultivation and to make collective efforts for their conservation in future.

This study has satisfied the need for understanding and documenting these indigenous knowledge based systems. With the Konyaks' positive participation, the results of the study provide the basis for further improvement of the local practices of conservation and use of plant genetic resources.

Participatory methodology for studying local plant genetic resources management

Participation was used as part of the study methodology with the following aims:
- to understand existing plant genetic resources conservation systems based on indigenous knowledge
- to increase awareness about the value of plant genetic resources
- to mobilize the community in the process
- to design the strategies for long-term plant genetic resources conservation
- to develop an effective framework for execution of these strategies.

To achieve participation for the first two steps, regular PRA techniques and tools like group discussions, interviews, village meetings were sufficient. However in the later stages a modified and more direct approach has been developed. Figure 23.1 shows the various steps followed and tools used in this approach.

Step I

As the initial phase of the process of participation, documentation of the indigenous knowledge (IK) associated with overall natural resource management and PGR conservation was initiated with the help of an interpreter. These preliminary inquiries helped to locate interested, fairly educated individuals from the community who worked in future as local researchers or

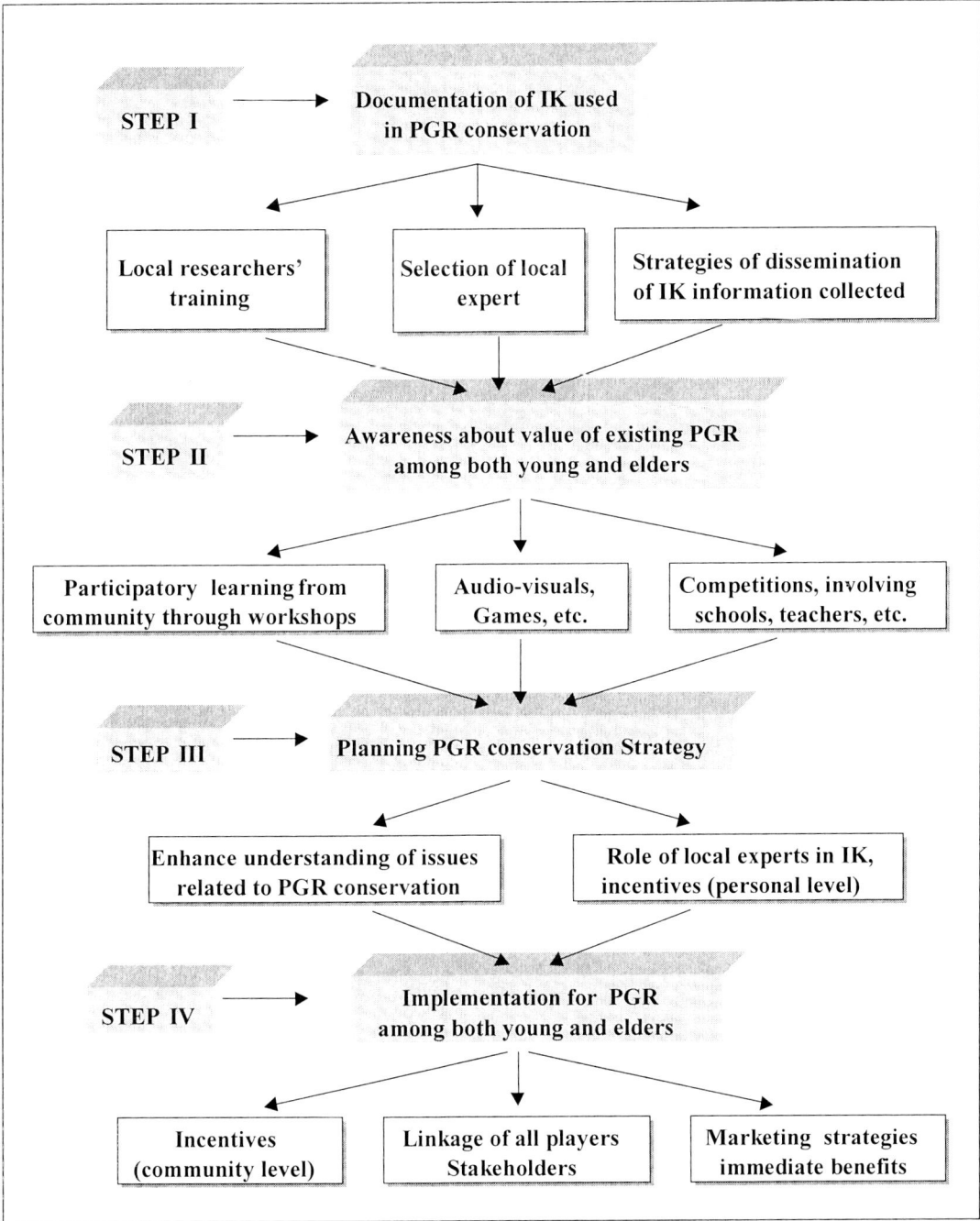

Fig. 23.1. Participatory approach for conservation research.

facilitators. The initial documentation process was instrumental in selecting the knowledgeable old people from the community as local experts. At the same time, with the participation of local researchers and experts, a strategy for dissemination of the information and indigenous knowledge was designed.

Step II
In the second phase, with the participation of trained local researchers/facilitators and experts, these planned strategies were implemented to create awareness about the value of existing PGR among the younger generation. The tools used to achieve participation at this level were learning and exchange of understanding about indigenous knowledge through interactive workshops, audio-visual shows, games, etc. To involve the children in the process, drawing and essay competitions were organized. In community-interactive workshops, schoolteachers were specially invited and trained to organize programmes for children.

Step III
With the participation, discussion and common consensus, a process of planning a PGR conservation strategy for the future was initiated. The community, AERF (an outside research organization) researchers, teachers, local researchers and local experts together categorized priority issues related to PGR conservation. The role of local experts and their IK was very important. At this stage all the players agreed to provide incentives to local experts, as the guardians of knowledge and helping to make use of the same today. During the process of initial research and documentation the incentive was in the form of financial/monetary provisions to a limited extent, provided by the research agency. However, in the process of PGR conservation at community level, it will be in the form of continuous benefits in kind as per the traditional systems.

Step IV
Implementation of the designed PGR conservation strategy is the final stage and obviously most difficult. It is necessary to provide some incentive to the community as a whole, but this is possible only through a long-term project until the community becomes self-sufficient and forms the beneficial mechanism of PGR conservation. Well-planned sustained marketing strategies for PGR could be one of the means to get immediate monetary gains. Establishment of a marketing set-up that covers a greater geographical area than a village is necessary. Again, involvement of stakeholders other than the community, such as local NGOs, government machinery, traders and buyers will be extremely important. Their linkage with the community and their awareness about issues related to PGR conservation are mandatory and possibly can be achieved through continuous interactive training. Of course it will be a long and continuous process.

Main elements in the participatory approach

Training of local facilitators
Local informants provided correct information, but little beyond that. To overcome this difficulty, we selected two Konyak young people with basic education and carried out two weeks of training sessions which included:
- prolonged discussions between researchers and the two local facilitators
- carrying out joint fieldwork in the Konyak area
- exposure visits to more distant areas such as Western Ghats.

A major aim of the training was to provide the trainee with an understanding of the concept of documenting indigenous knowledge, use and conservation of plant genetic resources. This included training in tools such as resource mapping, ranking of commodities available according to use, separation and listing of available plants and plant parts along with transect walks in groups. The exposure to culturally different areas moreover helped the local facilitators develop a more complete view of the outside world and provided the basis for an understanding of the value of their own culture and respect for local practices of plant genetic resources conservation and use.

After completing their training the two local facilitators were employed by the research project as local researchers (staff members for the project period) and assigned a role as the link between the research team and the local community. This has functioned successfully. As an example the local facilitators designed interactive programmes on their own for communities involved in the project with the aim of increasing the general awareness about plant genetic resources conservation and use. These two Konyak facilitators trained at least one young person with basic education in each of the nine villages included in the research project. The quality of selection and training of local facilitators is crucial for the success of community participation.

These facilitators were full staff members of the project for 2 years and were paid a monthly salary. They remained in touch with the community even after the project finished and are in the process of translating the final report into the common Konyak language. If we get support for the research in future they will work as regional coordinators rather than just as facilitators or researchers.

Stakeholder analysis and involvement

Stakeholders here are considered to be the community as a whole, government agencies and officers, NGOs working for rural development, education or for any other cause, outside research agencies and all others related to the issues and problems of PGR conservation and could be involved in future for resolving these issues. For example if we involve the government agencies in the process of a participatory-learning-awareness-generation process, it will be easier in the future for the community to use government machinery effectively in the community-managed programmes for PGR conservation.

In a remote area like Mon district the stakeholders involved in the process of plant genetic resources conservation issues and management strategies are comparatively limited. Individuals within the community are the major stakeholders. Within the close-knit structure of the community, different groups of individuals are responsible for a variety of functions. Strategies to achieve participation of each subgroup therefore differ largely. These subgroups are the elder generation of knowledge-holders, the present generation of practitioners which is engaged in contextual modifications within the system as needed, the younger generation which hardly accepts the value of culture and IK-based non-commercial resource management practices, and children. Women may be considered separately or as part of each subgroup. Village administrative machinery and government officials form another group of stakeholders and yet another group of stakeholders is a large number of outsiders or migrants, especially non-tribals. Their participation has not been considered important for plant genetic resources conservation, as they are not part of the traditional village structure. Immigrants are comparatively far away, they have their own belief in the modern development systems, and are mostly dependent on the govenment. In many cases they try to imply that instead of participatory mechanisms, government should take all the responsibility for development and conservation. Therefore, at least in this first effort of participatory planning, immigrants have not been considered among the stakeholders.

In our research we have concentrated on knowledge-holders, practitioners and the younger generation as well as children. Women formed the integral part of the total programme but were not considered separately except for some ranking exercises. Interactions with these subgroups using various tools like resource mapping and transect walks helped to understand the major issues related to plant genetic resources conservation and the need to work together as a community with other stakeholders. Participation in the initial stages was restricted to only two or three subgroups of stakeholders but as the work continued all the groups got involved and cooperated.

Participatory documentation workshops

Indigenous knowledge about and strategies for plant genetic resources conservation and use were collected from three villages from lower Konyak territory by carrying out participatory documentation workshops. The official village-level community organization is strong and is the main decision-making framework. The village administrative bodies, i.e. Village council and Village Development Board, therefore had to be involved in organizing the workshops. Local government officials, such as the 'extra additional collector' (EAC) were also involved from the beginning, which contributed to the success of the project. EAC is the State Government official appointed for the overall well-being and development of the area under his jurisdiction.

The trained Konyak facilitators identified local plant genetic resources experts from each village. These local experts functioned as facilitators during the participatory documentation workshops. The response from people during the participatory documentation workshops has been overwhelming and many of farmers emphasized the need for further training sessions for village elders/schoolteachers and other knowledgeable individuals from the villages.

These workshops not only helped to collect and analyze/interpret the data for assessment of available plant genetic resources but also proved important for focusing issues and problems associated with plant genetic resources conservation in today's context. These workshops were thus most successful for awareness-raising as well as for understanding the crucial role of resources in the livelihood of Konyaks.

During the participatory documentation workshops, the community expressed a need for improved marketing of home garden products. Many of the plants grown in the home gardens meet the needs of small townships like Mon and Aboi along with the household consumption. However, some local farmers indicated that if a better market framework were not established in the near future, cultivation of specific local varieties of beans and garlic may not be continued. If plant genetic resources conservation is directly linked with economic benefits, it is far easier to get continuous community participation in any programme.

Language and inter-village exchange of knowledge of plant genetic resources use

Language was a major barrier for us in the initial stages. The help of interpreters was crucial but time consuming. The Konyak tribe is subdivided into two cultural subgroups known as upper and lower Konyaks. These subgroups are located in high- and low-altitude areas. Each Konyak village has dialects which are significantly different from each other. Within the last 10-15 years, Konyaks of both the areas have selected one language (Wakching) as a common Konyak for communication among all the villages as well as for education.

We translated our first year's findings into this common Konyak with the help of our local researchers and experts. Distribution of these translated reports of three villages from the lower Konyak area to the villages from upper Konyak area generated tremendous enthusiasm about the whole research programme, and plant genetic resources and indigenous knowledge became the important topics of discussions in Village Development Board and Village Council meetings. Following dissemination of the information about plant genetic resources of three villages to four other villages, a type of competition has been created for participation in recording ethnobiological knowledge. Translation of the results from the first year into a local dialect and its distribution among villages therefore was the most successful tool used in our participatory approach. Each of the nine villages where we have conducted the research now has a copy of the research results and details of traditional practices as well as plant genetic resources around them. Such written records are helpful to establish clear rights of the community or particular village over the information produced in the report. The resource maps prepared by the local experts and village together helped to resolve certain resource area conflicts and a common consensus has been developed.

Achievements and constraints of the approach

Participation of a large part of the Konyak community was achieved within a very limited

time using this participatory approach. The described approach can be replicable in locations where the influence of the outside world on the traditional society is comparatively low and traditional systems of conservation are still working. The achievements include the following.
1. Documentation of plant genetic resources available in home gardens and *jhum* cultivation practices of Konyak Nagas within a 9-month period. Most important is the fact that documentation was done by local people and the outside research agency worked as a technical secretariat.
2. This approach and new tools like reports for a community in its own language helped to disseminate valuable indigenous knowledge based information in all nine project villages and initiated the process of comparison and exchange of traditional knowledge by the communities involved.
3. Awareness about the available natural resources and value for indigenous knowledge has been enhanced. Age-old traditional knowledge mechanisms have started functioning again because of increased awareness about the value of culture. The continuous interaction process helped to re-establish links between older and younger generations.
4. A process of formulating local initiative for future protection of plant genetic resources has started.
5. Large-scale motivation about the value of plant genetic resources and issues related to their conservation has started among all the stakeholders.
6. Capacity-building for research and action/implementation work has been achieved and local leadership for plant genetic resources conservation activities has been developed.

Such an approach has prepared the grounds for formulation of future resource management plans based on community and indigenous knowledge. This will make resource use more economically beneficial and sustainable.

Our approach was able to achieve active and positive participation with documenting plant genetic resources conservation and use and discussing plans for improvements. It is, however, very difficult for a research organization within a limited time to facilitate the required level of consensus among all stakeholders, which is required to develop and execute improvements in existing practices. The participatory approach clearly shows that plant genetic resources conservation action is important for the small farmer. Such understanding may in the future lead to sustainable farming.

A major difficulty of this approach is the question of providing incentives to individuals and a community as a whole. It is easy to provide monetary incentives to individuals who were directly employed as facilitators or local experts. However, long-term incentives in the form of regular programmes for village schools or equipment for the school and Village Development Board are acceptable non-monetary options. Deciding on such matters in a participatory manner is time-consuming and uses time otherwise spent on the actual documentation and action programme.

The poor level of communication and marketing linkages in the Konyak area are major constraints to enhanced valuation of local plant varieties which would provide incentives and benefits of the plant genetic resources conservation strategies. Therefore, dissemination of information not only in the local language but also for the government machinery is very important.

It was beyond the capacity of our research project to facilitate negotiation and involve government organizations in long-term participatory plant genetic resources conservation programmes.

Acknowledgements

Development and successful testing of this modified participatory approach could not have been possible without the perfect understanding and support of our Konyak Research Fellows Mr Loipong Konyak and Mr Pungchei Konyak. Valuable financial support by Earth Love Fund, UK for this documentation is greatly acknowledged. Thanks are due to Mr A. M. Gokhale, Chief Secretary, Government of Nagaland, and Mr R. Kevichusha, Team Leader, NEPED Project for encouragement and field-level support. Enthusiasm and participation of all the villagers from the nine project villages is highly appreciated and acknowledged. The help of colleagues from AERF-Swapna Prabhu and Aparna Watve is acknowledged.

24. Linking to community development: using participatory approaches to *in situ* conservation

P.V. Satheesh

Introduction

When you are dealing with something as important as rural development, genetic conservation, biodiversity and control over media – issues that are so grave in their import – why and how will you use participation? Is it at all necessary and/or prudent to push participation into such important issues? When we started an *in situ* genebank programme, we at the Deccan Development Society (DDS) constantly faced questions like this.

The DDS is not an academic or research institution. It is a small grassroots development organization working in the semi-arid tropics of Deccan in South India for over a decade and a half. It is devoted to reversing the degradation of the environment and people's livelihood systems through a string of activities, like permaculture, community grain fund, community green fund, community gene fund and collective cultivation through land lease. The society presently works with about 100 Dalit women's Sanghams (village-level voluntary associations of the poor) consisting of nearly 4500 members in 75 villages in the Zaheerabad area of Medak District, Andhra Pradesh State, India. These are at the lowest rung of the ladder in the Indian socioeconomic hierarchy and form the poorest section of the rural community. Though these women intuitively practise biodiversity and conservation of genetic resources as part of their survival strategy, they do not appear to be knowledgeable of ecological concepts. Or do they have any special clues?

When the DDS began a programme on *in situ* genetic conservation, there was total confusion. On one hand, the problem was how we could communicate these concepts to illiterate rural Dalit women and on the other hand, how we could learn about these concepts from them. What is the *lingua franca* we use for communication of these concepts – participatory tools and approaches? That was a familiar terrain for us because in DDS nothing can take place without the use of participatory tools. So why not use the same tools for this programme also?

Community Gene Fund

In 1995, DDS began an ambitious programme of retrieving more than 50 landraces that were under active cultivation in this region of Deccan as recently as 30 years ago. For us it was the beginning of the Community Gene Fund, a grandiose name to indicate an *in situ* genebank programme. For the Dalit women of the DDS family, it was just a Paata Pantala Programme, a programme related to traditional crops. Through the very act of naming, they had demystified the concept from the scientific jargon that surrounds it. The community Gene Programme, which was funded by the GTZ, made available to every partner woman farmer an amount of Rs. 3000 (approximately US $75) per acre as a loan. This money was used for repeated ploughing and manuring of the degraded lands, where cultivation of local crops would begin. The partner farmers would repay this money, in the form of a variety of seeds of traditional landraces from their farms, to their own autonomous village *Sanghams*. These seeds would form the village-level genebank and would be available for any farmer who wished to plant the seeds in her/his farm.

The core objective of the programme also evolved through a participatory analysis of the problem. A series of discussions, mapping of agricultural landscapes, time lines and matrices came in very handy to aid these analyses. Through these analyses, the Dalit women realized that the loss of their traditional crops was, in fact, the result of triple marginalization:
- marginalization of a range of millets (little millet, foxtail millet, finger millet, proso millet, pearl millet, kodo millet), and a range of pulses and grams which addressed the needs of the poor

- marginalization of lands that cannot produce commodity crops for the market but support a range of coarse cereals consumed by the poor
- marginalization of people who grew these kinds of crops. Dalit women who had the knowledge and capacity for the cultivation of these crops were automatically marginalized when their crops and lands were marginalized.

What is the principle we need to adopt for the retrieval of these landraces? The women felt that if conservation was the goal, the only path to pursue was to combine all elements in the triple marginalization and see whether a new synergy could emerge. Their hunch was to come true within 3 years. Within this period, about 59 *kharif* (the cropping season beginning in June) varieties were back in active cultivation in the fields. The process, which used a variety of participatory methods at all stages, gave us extremely rich insights into the cropping systems and rationale behind them, and farmers' perception of their crops. Participating in a study sponsored by the Using Diversity group of the International Development Research Centre (IDRC), we tried to find out how women farmers perceive biodiversity in their fields. The process and the results were an extraordinary learning experience for us.

The most exciting perception came in the classification of crops. The farmers called one set of crops *Satyam Pantalu* – Crops of Truth. These crops, they told us in a poetic description, survive on air. *Satyam Pantalu* are the winter crops like winter sorghum, safflower, chickpea, wheat, linseed and mustard (Fig. 24.1). None of them require any irrigation or rain. They grow on the residual subsoil moisture available in winter. Farmers of the Deccan who have always carried out their agriculture in the harshest of circumstances find these crops, which demand no inputs from them, are truly the Crops of Truth. Truth for them is the chance to survive without having to struggle.

This overarching world view which connects their belief systems and the harshness of their environments to their agriculture and the genetic properties of their crops is also evident in several other perceptions. For example, when they describe the properties of foxtail millet, the reference point for the women farmers of the DDS is an archetypal story from the Hindu epic Mahabharata. Pandavas, one of the clans in the epic, were exiled into isolation and forced to live in the forest for many years. During this period, there was a time when the hungry Pandavas could find nothing to eat but a few fistfuls of foxtail millet. They roasted the millet, rubbed it between their palms to dehusk it and blew the husk away. When the husk fell on the ground, it dramatically germinated and raised another crop. So the story

Fig. 24.1. Bringing back *Satyam Pantalu* – Crops of Truth into cultivation: A field of rabi (winter) crops seen to consist of winter sorghum, lentils, safflower, peas and mustard inter-cropped in the same field. This field actually hosts over 10 species and more than 16 varieties, all traditional landraces.

goes. That this amazing millet, even when roasted, can regerminate through its husk makes it a dream crop for the food-starved farmers of the semi-arid Deccan. Therefore it is also an unending crop (*Barkat Panta*).

Such extraordinary spiritual metaphorization of a hardy grain, which is the first to mature in the cropping cycle of the Deccan and answer the first pangs of hunger at the right time (*Maa Aakaliki Vastadi*), symbolizes their perceptions of crops and grains. This has been a wonderful source of rich learning for us. When the women did a matrix on the crops that they would like to retrieve, they listed certain landraces that they would like to continue growing for the reasons mentioned beside them.

- Soil types determine growing of certain crops: e.g. winter sorghum, lentils
- Crop duration are an influence: e.g. Gareeb Jonna, an early maturing variety
- Need to raise two or more crops during a year: e.g. green and black grams
- Need to make money: e.g. Bishop's weed
- Need for food: e.g. kharif sorghum
- Need for fodder – pod mix/straw/dry fodder: e.g. sorghum and pulses
- Need for thatching/fencing material: e.g. pigeonpea
- Need for fibre: e.g. amaranthus and sunhemp
- Need for vegetables: e.g. cucumber and cowpea
- Special foods for specific festivals: e.g. Pyalala Jonna (popping sorghum)
- Need for rejuvenating soil fertility and 'strength': e.g. niger
- Storability: e.g. foxtail millet
- Need for oil: e.g. safflower
- Need to prepare land for the next crop (green-manure crop): e.g. sunhemp
- Medicinal – for various ailments: e.g. mustard
- To ward off/reduce pest incidence: e.g. mustard, marigold
- Need for fuel: e.g. pigeonpea.

Through a series of such matrices we were able to discover the following practices:

- inter-cropping in a certain manner during *kharif* will reduce disease incidence in the *rabi* season, the second cropping season, which comes 6 months later
- traditional pest-control practices were scored for their efficacies
- soil-building crops of the region were inventoried
- various seed-sowing practices for different crops were listed
- rationale behind the traditional inter-cropping and mixed practices was explained
- nutrient needs of various crops and their inherent productivity values were listed.

Thus, a wide range of amazing agricultural practices and the farmers' perceptions enveloping such practices became the framework for the genetic conservation programme. The difference that this approach made to the programme was enormous. *In situ* conservation was no longer a scientific idea mooted by outside conservationists. It became a programme defined by insiders' perceptions, their needs and their understanding. This was the inherent strength of this programme.

At the end of a 3-year crop cycle, there was a participatory evaluation of the programme. A group of women farmers representing those who entered the programme at different stages – first year, second year and third year – used participatory tools to assess the gains they had made through the programme. Here is what they said through this matrix (Table 24.1).

As can be seen in this self-evaluation, the issues presented by the women as a rationale for their active defence of the diversity in their fields, range from the political to the practical. They gave the highly charged political question like local control over seeds as much importance as a practical consideration like the availability of a variety of fodders. They also placed a huge importance on crop diversity to fight the risks inherent in the dryland agriculture in a degraded, semi-arid region.

Table 24.1. Matrix of self-evaluation developed by women from 13 villages

Objectives	Rabi 95-96 to Rabi 96-97: 3 seasons	Kharif 95-96 to Rabi 96-97: 2 seasons	Rabi 96-97: 1 season
Fertility achieved with FYM application (%) [indicated by manure heaps]	75	50	–
Revival of lost landraces (%) [indicated by a field recreated with a variety of intercropping]	100	100	25
A variety of healthy foods to eat (%) [indicated by a picture of a women and a stove]	100	50	25
Leaf fall and green manure value (%) [shown by leaves]	75	50	25
Increased fodder varieties (%) [shown with samples of different fodder]	100	100	25
Increased friendship between women and seeds (%) [shown with a picture of a women and variety of seeds]	100	50	25
Outsiders approach them for seeds (%) [shown by many crops in one field]	25	–	25
Risk insurance with increased diversity (%) [shown by many crops in one hand]	75	50	25

Through the Gene Fund project period, such participatory exercises and discussions gave a forum for the women farmers not only to educate us in the DDS on the finer points in diversity and conservation but also to exchange the information among themselves and keep the exercise of mutual learning going.

Biodiversity Festival

At the end of the project, the women organized a biodiversity fair, which they christened as *Paata Pantala Panduga*, a Festival of Traditional Crops. This was an amazing setting in which the women used the traditional fair format (display rooms) to discuss a wide range of issues (see Box 24.1):

- the range of traditional landraces they had revived (seed room)
- the crop design on their lands and the rationale behind the design (map room)
- the ritual and belief systems that surround them (ritual room, Fig. 24.2)
- a variety of recipes that these grains can be used in (also the ritual room)
- the range of diverse grasses and fodder such agriculture produces (livestock room)
- the capacity of these farm wastes to build their homes, cover their roofs and to feed their cattle (also the livestock room).

The fair also helped to reassess and retrieve the biodiversity wealth within the community, and turned it into a cultural forum for the women to present their arguments in defence of diversity using an array of cultural idioms. They also converted the fair into a discussion forum to debate the arguments around the economics of a biodiversity-based production system. The powerful arguments that they marshalled on behalf of their own brand of agriculture won the

Box 24.1. A brief account of *Paata Pantala Pangdugal* (Festival of Traditional Crops).

The Seed room
This proudly exhibited the collection of the diversity of seeds in the region. Seeds of over 85 various crop species and subspecies were on display in this room. Alongside the seeds were various methods of storage in woven baskets, large earthen pots and other indigenous storage structures.

The Map room
Six most fascinating women among the DDS family, who have maintained a vibrant diversity on their farms, displayed the crop varieties they had planted this season. These women were present in this room to explain the reasons for planting a number of crop varieties, the design they have followed and the rationale behind those designs.

The Ritual room
In this room, six extraordinary agricultural festivals and rituals followed by the farmers in Deccan were recreated. It started with *Penta Pooja*, the worship of the compost heap, followed by *Chaviti Pooja*, the worship of farming tools that takes place on the most auspicious day of the Hindu days, namely *Ganesh Chavity*. The next recreation of the ritual was *Erokka Punnam*, which occurs just before the start of the agricultural season. On this day farmers reverently worship their most prized possession, the plough bullock. Then comes *Dasara* festival, in which the best seeds of the house/village are collected and given a germination test in an isolated place of the house called *Gattilu Koorchodam*. On the *Soonyam Pandugu* day, the fifth ritual recreated in the festival, farmers cook special foods and visit their farms which are in full bloom, sprouting fruit/grains of the diverse crops they grow. They make images of *Pandavas* in their farms, worship these images and then go around the farm offering food to pregnant earth mother and pray for her well-being. The last of the reconstructed rituals was the *Endlagatte Punnam*, the most delightful celebration of biodiversity in the Deccan. On this day farmers bring home a variety of fruit/grains of different crops growing on their farm and string them above the doors of their house. They first offer this pre-harvest of crops to *Ooredamma*, the Village Goddess. The ritual room also presented a variety of traditional recipes made out of traditional crops.

The Livestock room
Organized like a typical cattleshed to the last detail, replete with shepherd's high bed, this enclosure proudly exhibited the wide variety of local breeds of cattle, small ruminants like goats and sheep, and poultry. At the centre of this enclosure was a plot which reminded the visitors of the enormous variety of traditional fodder species still available in this region.

The Exchange room
Finally, the exchange room where all the women who attended the *jatra* (festival), came together to discuss their impressions of the *jatra*, the issues brought up in the *jatra* and the ways ahead. Their discussion was facilitated by a team of traditional farmers who helped the visiting groups of women farmers from the 75 *Sanghams* to get over their collective amnesia about the enormous wealth of traditional seeds and cropping systems they have in the region. They also exchanged their own perceptions about the diversity in their farms.

Fig. 24.2. The Ritual room.

day when they proved that their agriculture was nearly 40% more profitable than the modern agriculture, even in monetary terms.

The incredible energy and vibrancy that these exhibits, demonstrations and discussions produced was evident on the last day of the fair, when about 5000 women joined in a human chain around a specially grown diverse field at the centre of the fair venue. They took a collective pledge in ringing tones:

> "We will defend the diversity on our fields. We will enhance this diversity and bring back more traditional crops into active cultivation"

The *jatara* (festival) had opened up more possibilities for people in the area of participatory communities. An important tool used in the *jatara* was the videos produced by the community of the women themselves. They also filmed the entire *jatara*. What was the significance of the video at this grassroots level? This was an approach designed by the Deccan Development Society to celebrate the knowledge of illiterate local communities and provide them with a new medium for articulation.

Participatory video filming

In fact the effect of the Society "handing over the camera" to rural illiterate women was to highlight the participatory efforts of the organization. Since the majority of the women DDS works with are not literate, we thought that it was very important to give them skills in handling an alternative tool to negotiate with the outside world in place of literacy. This tool was intended to make these women take a critical look at the processes they were involved in and offer their critique through the video camera. What would not have been possible through traditional literacy methods became an exciting possibility through media literacy.

These groups of video women made an exciting film on the *jatara* and are in the process of sharing this experience with other women in the Society as well as the larger outside world.

Conclusion

In all these efforts, from creating a Gene Bank to a Grassroots Media, participation has been the underlying thread. Participation has been used for people's own interpretation of their context for genetic diversity and its revival, their strategies for its enhancement and maintenance, their methods for the propagation of this knowledge and awareness within and outside of their own communities. The rewards have been very rich.

25. Identifying and analyzing policy issues in plant genetic resources management: experiences using participatory approaches in Nepal

Devendra Gauchan, Anil Subedi and Pratap Shrestha

Introduction

International, national and local policy environments play important roles in the management, ownership and control over local plant genetic resources (PGR). Emerging international policies on PGR include intellectual property rights, farmers' rights, the rights and interests of indigenous local communities, access to genetic resources, benefit-sharing, and *sui generis* protection of plant varieties. In the last three decades, several international conventions and agreements have taken place worldwide to discuss and develop policy dimensions of PGR. The Convention on Biological Diversity (CBD) is a significant landmark among international agreements, since it incorporates for the first time the principles of ethics and equity in both access to genetic resources and sharing of benefits (Swaminathan 1998). Subsequent to the implementation of CBD, the World Trade Organisation (WTO) was created in 1994 and Trade-Related Aspects of Intellectual Property Rights (TRIPS) came into force in 1995. With these undertakings, several policy issues related to PGR management have emerged worldwide. According to TRIPS, countries have to "provide for the protection of plant varieties either by patent or by an effective *sui generis* system or by any combination thereof". The TRIPS agreement sets out compulsory, uniform standards for intellectual property protection throughout the world by patents, trademarks and copyrights. To date, however, only the International Union for the Protection of New Varieties of Plants (UPOV) provides a workable example of an effective *sui generis* system. The liberalization of world trade, the new international trade regulations and Intellectual Property Rights (IPR) formulated by WTO, genetic engineering, patenting and terminator technologies make it possible for powerful multi-national commercial interests to exert a decisive influence on access, control and ownership of PGR. This has resulted in a threat to local biodiversity, farmer's income, food security and food safety.

Nepal has taken part in a number of international agreements and conventions and has been actively involved in the discussions leading to them in various preparatory meetings. Nepal is a signatory of the CBD held in 1992, but further initiatives in implementing the recommendations of the conventions have been greatly lacking. Nepal's proposed entry to the WTO and enforcement of TRIPS in WTO have brought new policy challenges and issues that need increasing concern and internal preparation (Gauchan *et al.* 1999). At present, there are no policies, acts and laws for the sustainable utilization and conservation of agrobiodiversity on-farm (Rijal *et al.* 1997; Pant 1998; Gautam *et al.* 1999). In the absence of specific policies on farmers' rights and *sui generis* systems in place, the country lacks programmes for effective management, utilization and conservation of PGR. Present policy on biodiversity is more focused on forest resources including wildlife than on the overall genetic diversity encompassing agricultural crops.

An enabling policy environment – especially the incentives it generates through price, credit, technology, institutions and regulatory framework – plays a pivotal role in sustainable harnessing, value addition and conservation of PGR (Gauchan *et al.* 1999). Activities to support the creation of policy measures include the identification of policy constraints and incentives (legal, institutional, market and non-market), and the proposals of policy changes detrimental to landrace diversity (Jarvis and Hodgkin 1998). Enhancing the knowledge base and understanding of policy issues among policy-makers as well as other stakeholders are necessary for effective management and utilization of plant genetic resources. However, there are very limited policy studies so far conducted in Nepal that provide examples of appropriate methodological approaches useful in the analysis of policy issues related to PGR and thereby facilitate policy adoption or changes.

Most of the policy analysis studies carried out so far are at the macro or national level, which are not appropriately and adequately integrated with the micro or grassroots policy issues and constraints. This article attempts to describe and discuss the participatory policy analysis methods useful in understanding policy analysis linking different stakeholders at different levels.

Methodological process
A participatory approach to policy research is a new and evolving field. Unlike conventional policy analysis, it involves a number of key stakeholders at different levels and methodological processes/steps for such analysis are identified and determined in the process. In a recent World Bank study, a multiple stakeholder analysis was employed for the Bank-funded policy study using the stakeholders' own indicators (Robb 1998). Its application in other fields, however, is a subject for further research. Well-established participatory policy analysis methods suitable for different subject areas and environments are yet to emerge.

In Nepal, the Natural Resource and Related Resource Management Project, implemented jointly by the Ministry of Agriculture (MoA) and the Winrock International Nepal, adopted participatory methods in policy analysis. This mainly included Participatory Rural Appraisal (PRA) tools, for example seasonal food and venn diagrams, in analyzing policy issues for a Tarai Food Grain Study (Gills 1992, 1998). These tools have been found useful in hypothesis-testing and policy analysis in agricultural resource management.

The participatory approach elaborated here is based on a study undertaken by Gauchan *et al.* (1999) in conjunction with NARC/LI-BIRD/IPGRI global *in situ* agrobiodiversity project in Nepal. The study employed participatory methods and involved a number of stakeholders at different levels (macro, micro and intermediate levels) to identify policy gaps, constraints and to analyze implications of national policies for agrobiodiversity conservation and utilization in Nepal. The steps and outputs of the methodological process are inter-linked and sequential: the first output leads to the second step and the second output leads to the third output and so on. The methodological processes used are described below.

In-house discussion among team members
First, the research team members, which included researchers from NARC, LI-BIRD and IPGRI, held a series of brainstorming discussions to identify broad researchable policy issues, various relevant stakeholders, and to conceptualize and develop a methodological framework for the study. The methodological steps described here were identified and agreed during those discussions. The in-house discussions also were useful in establishing a common understanding of the policy issues among the researchers.

Review of existing policies
This employed reviews of relevant literatures, especially policy and planning documents, as well as discussions with relevant experts. This helped researchers to update their understanding of the policy-related knowledge/information, identify policy gaps, refine methods and name lists of stakeholders identified earlier in step 1 for policy research.

Discussions with key representatives of different institutions
In order to further verify and reinforce the findings of the literature review, and identify specific policy issues and perception/interpretation gaps, key personnel of the concerned institutions were individually interviewed using a checklist. These institutions included the Ministry of Agriculture, National Planning Commission, Nepal Agricultural Research Council and the Institute of Agriculture and Animal Science (an academic institution). The personnel interviewed were

directly involved in policy analysis and formulation, and programme design and implementation related to agricultural research and development and to the formal education sector.

Discussions with farmers and traders
Focus group discussions (FGD) and key informant interviews were held with farmers' groups and traders in two *in situ* project ecosites, namely Begnas in Kaski district and Kachorwa in Bara district, to identify microlevel policy perceptions, implementation gaps, and specific policy incentives and disincentives at the community level.

Analysis and synthesis of the findings
Information generated from various steps during policy research was analyzed and synthesized through interaction and problem-causal diagramming drawn by the research team members. The synthesized information has been documented and presented to a wider audience involving policy-makers and planners for facilitating policy changes.

Findings of the participatory policy analysis
Policies are mainly of two types: economic and legislative. Economic policies include courses of action taken by the government in influencing economic environment of the farming community. Pricing, subsidy and credit policies are some of the examples of economic policies. Legislative policies are legal and regulatory frameworks of the government, which influence management of PGR at the national and local levels. These specific policy types and issues were studied and analyzed in relation to *in situ* agrobiodiversity conservation and utilization. The outcome of the policy analysis is briefly highlighted in Table 25.1 and discussed below.

Economic policies
Economic and agricultural policies play a major role in PGR management through farmers' decisions on whether or not to replace traditional landraces with modern varieties. Such policies can either revolve around agricultural practices directly, or they can shape the macroeconomic conditions, which determine the marketability of particular varieties. The notion that "economic benefits can be derived only from the promotion of modern varieties/technologies" is still the guiding philosophy in the policy formulation in Nepal.

Many of the present policies, formulated and implemented for agricultural development (e.g. credit, price, research, extension, etc.), have favoured production and marketing of modern varieties (MVs) of major crops. They have, therefore, either provided disincentives or no incentives in the conservation and utilization of minor crop and landrace diversity. Similarly, these policies do not provide any incentives for small-scale subsistence producers living in marginal environments who produce a small marketable surplus using low-input systems and traditional varieties.

Policies to multiply, certify, control quality and seed distribution of minor crops and landraces are lacking. Despite the large volume (almost 90%) of informal seed supply systems and their contribution to overall crop production and food security of small farmers, they do not receive any form of policy support from the government (Sthapit *et al.* 1996). Similarly, as elsewhere in the world, the formal education system in agriculture in Nepal is primarily geared toward imparting knowledge, skills and attitudes on the cultivation and promotion of modern varieties and technologies.

Legislative policies
The study revealed that currently Nepal lacks policies/legislative measures on on-farm conservation of crop genetic resources, particularly for threatened plant genes and valuable traits. It also lacks policies on farmers' rights and effective *sui generis* systems for sustainable

Table 25.1. Policy issues, gaps and constraints for agrobiodiversity conservation and utilization

Policy types	Specific policy gaps/ constraints and implications of policy for agrobiodiversity conservation, value addition and sustainable utilization on-farm
Agrobiodiversity policy/Act	• No separate agrobiodiversity Acts/ conservation policies • No policy or Act to conserve valuable plant traits and threatened genes • Lack of policy on effective *sui generis* and farmers' rights to recognize and reward farming communities for their knowledge and innovation
Germplasm exchange/Trade policy	• No one-window policies for the trade and transit (export and import) of seeds and plant materials in Nepal
Research policy	• Only general policy for crop improvement • Only conventional plant breeding methods are common • Limited research on minor crops and landraces
Agriculture extension policy	• Technology dissemination focused on major crops and modern varieties • No extension advice and inputs for the promotion of landraces
Marketing policy	• Lack of policies on value addition and marketing support (e.g. linking market networks and market facilities) for local crops and landraces
Regulatory frameworks	• Present variety release and seed regulatory framework require strict uniformity, quality standards and distinctness • No legislation and support systems (certification and quality control) for seed multiplication of landraces and minor crops
Education policy	• No policies to incorporate curricula, text books and teaching programmes in agrobiodiversity aspects in the university and extension programmes as in-service training, short courses, undergraduate and graduate training
Price policy	• No price support policy for minor crops and landraces
Credit policy	• Credit policy only for commercial production and profitable crops • Lack policies to finance credits for agrobiodiversity conservation purposes
Subsidy policy	• Input and credit subsidies are mainly directed to modern varieties • Food subsidies in remote areas have discouraged production of local crops and landraces and crops under threat or erosion • No subsidy policy for the promotion of minor crops and landraces

Source: Gauchan *et al.* (1999).

utilization, conservation and equitable sharing of benefits arising from the use of PGR.

Present varietal development and release processes are dominated by formal systems which provide little option for farmers to choose diverse cultivars, access to seeds, increase genetic diversity of the crops and add value. In addition, existing regulatory frameworks do not provide incentives for the rapid promotion of farmers' varieties.

Gaps in policy perception and linkage

Presently, agricultural policies are formulated by the policy-makers at the macro level without analyzing their relations and consequences on the micro level of on-farm management and utilization of genetic diversity. In addition, there are gaps in policy perceptions (interpretation) and implementation at the micro level among local traders and farming community. Lack of integration of macro level policy with micro level issues, users are less aware of policy incentives/ disincentives at the field level, while policy-makers are less informed about policy constraints and gaps in the implementation of the programme. This has resulted in the blockage of the two-way flow of information. Furthermore, policy-makers are less informed and aware of the potential benefits of plant genetic resources for food security and livelihood of people.

Facilitating policy changes

Awareness and advocacy to policy-makers and planners have to be attempted on policy gaps, disincentives, constraints and their implications at the programme planning and implementation level. Good policy always depends on good information, and this is particularly true for crop genetic resources (Tripp and Heide 1996). Good policy information and analysis are required to bring attitudinal changes among policy-makers and to enable them to make informed decisions. Although the final results in terms of policy influence are never fully predictable, identifying particular policy issues and tailoring the selection of methods are very important. There will, however, be a greater tendency than in conventional approaches toward having to deal with multiple stakeholders, expressing diverse interests through varying objectives.

It takes time to see changes/impacts due to better provision of policy information and/or to policy changes. The methodological approach employed and the information generated through policy study can be used to facilitate policy changes for PGR management. One of the possible ways envisioned in this aspect is organization of a policy workshop where research findings can be presented to policy-makers for informed decision-making. The findings of this policy study were presented in a Global *In situ* Agrobiodiversity Conservation Meeting at Pokhara, Nepal (5-12 July 1999). A number of policy issues have surfaced among various stakeholders, and it is hoped that the attention of the policy-makers and planners will be drawn to help influence policy changes in the desired direction. The study also has pointed out areas where policy interventions are required for sustainable PGR management.

Lessons learned

Policy research is a complex process that involves analysis of more than a set of goals and procedures and their interactions and impacts on multiple actors. Participatory policy research analysis is a process that takes place at different levels and with different actors. Policy-based research often involves a number of key stakeholders before the procedures and steps of the study are finalized. It helps create dialogue and link different stakeholders such as policy-makers, planners, research and development workers, and users as well as the external donor community (both formal and informal sectors) in identification, analysis and facilitation of the issues and options for policy changes.

The present methodological approach was useful for understanding and identifying policy issues, gaps, constraints and disincentives faced by different stakeholders and their differential perceptions/interpretations including gaps in implementation at various levels. The process/ steps followed helped to refine research methods, prioritize issues and verify findings on the specific policy issues. Since stakeholders were consulted separately, it was possible to get the individual views and perceptions of the various actors. In addition to identifying and analyzing various policy issues, this method has been found useful in the identification and sensitization of relevant stakeholders.

Owing to lack of time and resource constraints, it was not possible to undertake an in-depth follow-up study to understand linkages of macro policy issues with micro level user-perceived policy gaps. Similarly, a face-to-face dialogue among policy-makers, planners, research and development workers and users could not be undertaken. Participatory policy workshops linking multiple stakeholders (e.g. policy-makers, planners, research and development workers, users and donors) are proposed in the future to create dialogue and to analyze and integrate micro with macro level policy issues for facilitating policy changes through informed decision-making on policy disincentives and gaps.

Within the given time and resources, the study interacted with a small group of selected stakeholders. To make policy analysis more representative and to get more conclusive findings, consultations with different resource categories of farmers (i.e. commercial, subsistence producers, resource-poor), traders (i.e. wholesalers, middlemen and retailers) and diverse representatives of various institutions are required. Similarly, to study long-term policy implications, in-depth temporal case studies (involving continuous monitoring of policy implications) should be undertaken at various levels. Finally, there is a need to integrate policy analysis done at different (micro, macro and intermediate) levels and to link them in order to minimize policy gaps and perceptions.

References

Gauchan, D., A. Subedi, S.N. Vaidya, M.P. Upadhaya, B.K. Baniya, D.K. Rijal and P. Chaudhari. 1999: National Policy and its Implication for Agrobiodiversity Conservation and utilization in Nepal. Paper presented in "Third Global Participant Meeting on *In situ* Agrobiodiversity Conservation On-farm" held in Pokhara, Nepal, July 5-12, 1999.

Gautam, J.C., J.N. Thapaliya and H. Bimb. 1999. Policy on Agrobiodiversity in Nepalese National Context. Paper presented in National Conference on Wild Relatives of Cultivated Plants in Nepal, June 2-4, 1999. Organized by Green Energy Mission/Nepal.

Gills, G. 1992. Policy analysis for agricultural resources management in Nepal. A comparison of conventional and participatory approaches. HMG/Winrock International, Kathmandu, Nepal.

Gills, G. 1998. Using PRA for agricultural policy analysis in Nepal. The Tarai Research Network Food Grain Study. Holland and Black Burry, eds. "Whose Voice" Participatory Research and Policy Change.

Jarvis, Devra I. and Toby Hodgkin, eds. 1998. Strengthening the Scientific Basis of *In situ* Conservation of Agricultural Biodiversity On-farm. Options for Data Collecting and Analysis. IPGRI, Rome, Italy.

Pant, R. 1998. Addressing Broader Issues of Biodiversity /Agrobiodiversity Conservation within the Framework of National Policy: A Case from Nepal. *In* Managing Agrobiodiversity: Farmers' Changing Perspectives and Institutional Responses in the Hindu Kush-Himalayan Region (Pratap and B.R. Sthapit, eds.). ICIMOD/IPGRI.

Rijal, D.K., R.B. Rana and K.D. Joshi. 1997. Conservation of Non-Renewable Plant Resources: Problems and Prospects of agrobiodiversity conservation in Nepal. Paper presented for the national seminar on the "The National Biodiversity Action Plan" held on 24th October, 1997, Organized by RESOURCES NEPAL, Kathmandu, Nepal.

Robb, C. 1998. Participatory Policy Analysis (PPA): A review of World Bank's experience. Holland and Black Burry, eds. "Whose Voice" Participatory Research and Policy Change.

Sthapit, B.R., D. Gauchan and R.B. Rana. 1996. Rethinking in Research: Participatory Plant Breeding in Outreach Research Sites. *In* Proceeding of the Third National Workshop on Outreach Research, held on 23-25 May, 1996, Outreach Research Division, Nepal Agricultural Research Council, Khumaltar, Nepal.

Swaminathan, M.S. 1998. Farmers' Rights and Plant Genetic Resources. Biotechnology and Development Monitor No. 36. Various Options for *sui generis* rights systems. University of Amsterdam.

Tripp, R. and W. Heide. 1996. The erosion of crop genetic diversity: Challenges, strategies and uncertainties. ODI Natural Resources Perspectives, No. 7, March 1996. ODI, London.

Acronyms

ACRA	Asociación de Cooperación en Africa y America Latina
AERF	Applied Environmental Research Foundation, India
AGRUCO	Agroecología Universidad Cochabamba
BDI	Institute of Biodiversity, Ethiopia
BIOSOMA	Biodiversidad, Sostenibilidad y Medio Ambiente
BJRI	Bangladesh Jute Research Institute
CARE	A confederation of relief agencies
CBDC	Community Biodiversity Development and Conservation Project
CDR	Centre for Development Research, Denmark
CEDAGRO	Centro de Desarrollo Agropecuario
CENDA	Centro de Comunicación y Desarrollo Andino
CESA	Centro de Servicios Agropecuarios
CGIAR	Consultative Group on International Agricultural Research
CGN	Centre for Genetic Resources, Netherlands
CIAT	International Centre for Tropical Agriculture, Colombia – CGIAR
CIMMYT	International Maize and Wheat Improvement Centre, Mexico – CGIAR
CINS	Cooperazione Italiana Nord-Sud (an Italian NGO)
CIP	International Potato Centre, Peru – CGIAR
CNS	Consejo Nacional de Semillas
COPLA	Centro Orureño de Planificación
CPA	Central de Productores Agropecuarios
CPRO-DLO	Centre for Plant Breeding and Reproduction Research, The Netherlands
DDS	Deccan Development Society, India
DFID	Department for International Development, UK
DSE	German Foundation for International Development, Germany
FAO	Food and Agriculture Organization of the United Nations
FCH	Fundación Contra el Hambre
GARDP	Gulmi Arghakhanchi Rural Development Project, Nepal
GEF	Global Environmental Facility
GTZ	Deutsche Gesellschaft fur technische Zusammenarbeit, Germany
IARC	International Agricultural Research Centre
IBCR	Institute of Biodiversity Conservation and Research, Ethiopia
ICIMOD	International Centre for Integrated Mountain Development, Nepal
ICRISAT	International Crops Research Institute for the Semi-Arid Tropics, India – CGIAR
IDRC	International Development Research Centre, Canada
IDS	Integrated Development Systems, UK
IIED	International Institute for Environment and Development, UK
IITA	International Institute for Tropical Agriculture, Nigeria – CGIAR
INGER	International Network for Genetic Resources

IPGRI	International Plant Genetic Resources Institute, Rome – CGIAR
IRRI	International Rice Research Institute, Philippines – CGIAR
ISNAR	International Service for National Agricultural Research, The Netherlands – CGIAR
IUCN	International Union for Conservation of Nature and Natural Resources
LARC	Lumle Agricultural Research Centre, Nepal
LI-BIRD	Local Initiatives for Biodiversity, Research and Development, Nepal
MASIPAG	Farmer'Scientist Partnership for Development, Philippines
MDFSRDI	Mekong Delta Farming Systems Research and Development Institute, Vietnam
MYRADA	An NGO in south India
NARC	Nepal Agricultural Research Council, Nepal
NARS	National Agricultural Research System
NCPGR	National Committee on Plant Genetic Resources
NGOs	Non-Government Organizations
ODI	Overseas Development Institute, UK
PESEM	A seed-production enterprise in the Andes
PGRC	Plant Genetic Resources Center, Ethiopia
PIABS	Proyecto de Investigación Agraria Bautista Saavedra
PLAN	An international NGO
PROINPA	Promoción e Investigación de Productos Andinos
PROSUKO	Programa Interinstitucional Sukakollo
SEMTA	Servicios Múltiples de Tecnologías Apropiados, Wiñaymarka
SENDA	An NGO in Latin America
SGRP	System-wide Genetic Resources Programme – CGIAR
SoS	Seeds of Survival Programme
SWP–PRGA	System Wide Program – Participatory Research and Gender Analysis
UBINIG	Policy Research for Development Alternative, Bangladesh
UNCED	United Nations Conference on Environment and Development
UNDP	United Nations Development Programme
UNEP	United Nations Environment Programme
UPOV	Union for the Protection of New Varieties of Plants
UPWARD	Users' Perspective within Agriculture and Rural Development, Philippines
USAID	United States Agency for International Development
USC/C	Unitarian Services Committee of Canada
UTO	Universidad Técnica de Oruro
WRI	World Resource Institute, USA
WTO	World Trade Organisation

Glossary

Accession – Plant sample, variety or population held in a genebank or breeding programme for conservation or use.

Adaptation – The process by which individuals (or parts of individuals), populations or species change in form or function in such a way as to better survive under given environmental conditions.

Agenda 21 – A comprehensive action plan on the environment adopted at the United Nations Conference on Environment and Development in Rio de Janeiro, Brazil in June 1992.

Agro-ecosystem – An ecological system modified by people to produce food and other products for human use.

Biodiversity – The total variability within and among species of all living organisms and their habitats.

Biogeography – The scientific study of the geographic distribution of organisms.

Centre of diversity – The geographic region in which the greatest variability of a crop occurs.

CGIAR – The Consultative Group on International Agricultural Research, an association of private and public donor agencies which support the work of sixteen international agricultural research centres.

Character – An attribute of an organism resulting from the interaction of a gene or genes with the environment.

Characterization – Determination of the structural or functional attributes of a living plant.

Community Biodiversity Register (CBR) – A register maintained by a local community to record the local existence of and uses of biodiversity harvested as food or for other human uses. CBR is used to monitor the status of genetic erosion of agrobiodiversity at a local scale.

Community seed bank – Terminology used in this bulletin to distinguish a small community-level seed storage where seeds and planting materials are conserved under local conditions for short-term future need.

Conservation – The management of human use of the biosphere so that it may yield the greatest sustainable benefit to current generations while maintaining its potential to meet the needs and aspirations of future generations. Thus conservation is positive, embracing preservation, maintenance, sustainable utilization, restoration and enhancement of the natural environment.

Convention on Biological Diversity – A legally binding international agreement for the conservation and sustainable use of biodiversity. Its final text was adopted in Nairobi on 22 May 1992. It was signed by over 150 countries at the UN Conference on Environment and Development in Rio de Janeiro, Brazil, in June 1992 and was ratified by 128 governments as of October 1995. The convention came into force on 29 December 1994. The US had not ratified it as of early 2000.

Cultivar – A cultivated variety of a domesticated crop plant. Syn. Variety

Diversity block – A demonstration block of local diversity collected during the diversity fair to raise public awareness as well as to multiply seeds for exchange.

Diversity fair – A fair of local crops or livestock diversity organized by a local organization or genebank with the objectives of sensitizing the local community, locating and identifying custodians of biodiversity, and collecting genetic resources and associated knowledge to promote the exchange of materials and information.

Diversity kit – A material kit containing seeds and/or planting materials of a large number of crops and crop species distributed to farmers with the objective of increasing on-farm biodiversity. It is a simple way of deploying diversity in the farmers' fields by distribution of local crop diversity through grassroots organizations.

Ecosystem – The dynamic complex of microorganism, plant and animal including human communities and their non-living environment, interacting as a functional unit.

Emasculation – Removal of the anthers from a bud or flower before pollen is shed; removal of male flower parts or rendering them non-functional; a preliminary step in crossing to prevent self-pollination.

Empowerment – Empowerment is about people – both women and men – taking control over their lives: setting their own agendas, gaining skills, increasing self-confidence, solving problems and developing self-reliance. It is both a process and an outcome.

Endemic – Restricted to a specified region or locality.

Ethnobotany – A branch of science dealing with the folklore knowledge of plants.

Evolutionary breeding – Breeding procedure in which the variety is developed from an unselected progeny of a cross, or multiple crosses, that have undergone evolutionary changes (i.e. adaptation to changing local conditions).

***Ex situ* conservation** – Literally conservation 'off site'. Conservation of a plant outside of its original or natural habitats, e.g. in a genebank, botanical garden or field genebank and stored as seed, tissue, entire plant or pollen.

F_1 generation – The first filial generation, usually the hybrid between two homozygous parental types.

$F_1, F_2, F_3,$ etc. – notations that designate the first filial generation, the second filial generation, etc. after a cross.

Farmer Network Analysis (FNA) – A farmer network in a social system refers to the inter-relationship among a set of individuals for seeking information and genetic resources and exchange.

Farmers' Rights – In 1985, the UN Food and Agriculture Organization (FAO) Commission on PGR (now FAO Commission on Genetic Resources for Food and Agriculture) introduced the principle of Farmers' Rights. The FAO's International Undertaking (IU) on PGR was amended in 1991 to include Farmers' Rights and the focus is on national systems to implement rights. The amendments recognize farmers as past, present and future *in situ* agricultural innovators who collectively have conserved and developed agricultural genetic resources around the world. Farmers are recognized as innovators entitled to intellectual integrity and to compensation whenever their innovations are commercialized. Agenda 21 and the Biological Convention also have adopted the principle of Farmers' Rights.

Farming system – All the elements of a farm which interact as a system, including people, crops, livestock, other vegetation, wild life, the environment and the social, economic and ecological interactions between them.

Farming systems research – Adaptive research in which multidisciplinary teams are involved to assess the technological needs of research domains, and to target technology development to specific domains. Technologies are verified on-farm by researcher-managed plots in farmers' fields rather than in research stations, taking a farming system perspective.

Focus Group Discussion (FGD) – A group interview with a selected/focused group of people facilitated by an investigator. The discussion involves a relatively small number (5-15) of people and focuses on a specific problem or area.

Formal sector – General term used in this bulletin to refer to genebanks, professional researchers and plant breeders of governments and commercial companies.

Gender – Gender is a culturally specific set of characteristics that identifies the social behaviour of women and men and the relationship between them. Gender refers to social differences, as opposed to biological ones, between women and men that have been learned, are changeable over time and vary widely both within and between cultures.

Gender analysis – Gender analysis is the systematic examination of the roles, relationships and processes between women and men in all societies, focusing on imbalances in power, wealth and workload. Gender analysis also can include the examination of the multiple ways in which women and men, as social actors, engage in strategies to transform existing roles, relationships and processes in their own interest and in the interest of others.

Gender-disaggregation – Gender-disaggregation entails the collection and separation of data and statistical information by gender to enable comparative analysis/gender analysis (includes sampling of both women and men).

Gene – The functional unit of heredity. A gene is a section of DNA that codes for a specific biochemical function in a living organism in a laboratory.

Genebank – Facility where germplasm is stored or maintained in the form of seed, pollen or *in vitro* culture, or in the case of a field genebank, as plants growing in the field.

Geneflow – The transfer of genes from one breeding population to another as a result of migration, mating and gene exchange which may result in changes in gene frequency.

General Agreement on Tariffs and Trade (GATT) – GATT was established in 1947 and grew from a club of 23 industralized nations to an agreement between 115 signatory states. Following the Uruguay Round of negotiations (concluded in 1994), GATT came under the management of the multilateral WTO (World Trade Organization) on 1 January 1995. The Uruguay Round included an agreement on intellectual property as a trade issue, known as TRIPS (see below).

Genetic diversity – The genetic variability within a genus, species and variety; variation in the genetic composition of individuals within or among species.

Genetic erosion – Loss of genetic diversity between and within populations of the same species or the loss of entire species (e.g. wild relatives) over time or reduction of the genetic base of a species due to human intervention, environmental changes, etc.

Genetic resources – Genetic materials of plants, animals and other organisms which are of value as a resource from a genetic point of view for present and future generations of people.

Genetics – The science dealing with heredity.

Genotype – The genetic constitution of an organism; or a group of organisms with the same genetic constitutions.

Germplasm – A set of genotypes that may be conserved or used; synonymous with genetic resource.

Grassroots – Organizations or movements of people or society at a local level.

GxE – A classic approach in a formal breeding programme is to focus on genetic yield potential, minimizing genotype-by-environment (GxE) interactions, and seeking wide adaptation as a means to maximize programme impact. This usually assumes that variance in E (both in terms of agro-ecology and management) is minimized, and thus requires optimal conditions or high inputs. Not all farmers or regions are able to do this. For all of these cases, GxE will have to be exploited and not minimized to achieve good local adaptation. Literally it means to identify the relatively best-performing genotypes for specific niches (E). GxE interaction exists at various levels such as at the genotypic, QTL and phenotypic levels. Four types of GxE can be identified: (1) no GXE interaction, (2) heterogeneity, (3) crossover interaction, and (4) combined GxE interaction. Only repeatable GxE interaction is useful. The GxE component in an analysis of variance for a certain crop character (e.g. yield) can be separated into GxTemperature (e.g. cold-tolerant rice), GxRainfall (e.g. barley), Gx (waterlogged soil), etc.

Glossary

Habitat – A specific place that is occupied by an organism or community, and where interactions with other organisms and the environment occur.

Heritability – Capability of being inherited; that portion of the observed variance in a progeny that is inherited.

Home garden – Traditional farming practice around the homestead. Home gardens are land-use systems which involve the management of multipurpose trees, shrubs, annual and perennial arable crops, climbers, herbs, species, medicinal plants, fish ponds and animals on the same land unit, in a spatial arrangement or a temporal sequence. A home garden is a small-scale supplementary food-production system by and for household members that mimics the natural multilayered complex ecosystem. As home gardens display some characteristics of agroforestry systems it also can be considered as an agro-silvo-pastoral agroforestry system.

In situ **conservation** – Literally 'on-site' conservation; conservation of plants or animals in the areas where they developed their distinctive properties, i.e. in the wild or in farmers' fields.

In situ **documentation** – A process of documenting information related to plant genetic resources in their own habitats. In the case of agricultural crops, the documentation process includes four components of management: (1) locating diversity and site, (2) the farmers' system, (3) the seed-supply system, and (4) the monitoring system.

Indigenous knowledge (IK) – Knowledge that develops in a particular area and accumulates over time through being handed down from generation to generation.

Informal Research and Development (IRD) – IRD is an informal and simple method of testing, choosing and multiplying seeds of choice for development. The main purpose of IRD is to overcome the limitation of poor access by farmers to new crop cultivars that exist in the conventional research and extension systems.

Informal system – General term used in this bulletin to refer to farmers, their organizations and other CBOs/NGOs in plant genetic resources conservation and management.

Intellectual Property Rights (IPR) – Laws that grant monopoly rights to those who create ideas or knowledge. There are five major forms of IPR: patents, plant breeders' rights, copyright, trademarks and trade secrets.

Intensive data plot (IDP) – IDP is a participatory tool employed to understand the farm economics and farmer-managed practices at household level on a regular basis. All natural, socioeconomic and human-managed factors of the plots and households are measured and are related to the existing crop biology.

Labour calendar – Labour calendar is a simple tool that illustrates labour tasks performed throughout the agricultural season. This tool is especially useful for showing labour supply situations, critical peak period for labour supply and for illustrating gender-differentiated responsibilities and management of crops.

Landraces – Farmer-developed varieties of crop plants that are heterogeneous, adapted to local environmental conditions and have their own local names. In other words, landraces are farmers' varieties which have not been improved by formal or private/NGO breeding programmes. Modern cultivars can be grown by farmers and over a period of time, especially when self-seed is used and selection is practised, can 'evolve' into a landrace.

Local crop development – The continuous and dynamic process of maintenance, development and adaptation of germplasm to the environment, local agro-ecological production conditions, and specific household needs related to social differentiation, gender and ethnicity.

Mass selection – A system of breeding in which seed from individual plants selected positively or negatively on the basis of phenotype is composited and used to grow the next generation.

Matrix ranking – Matrix ranking involves asking farmers to evaluate each technology or variety of interest with respect to a number of criteria that are specified before ranking begins. For example, three tree species might each be evaluated for six criteria: (1) fodder value, (2) rate of growth, (3) usefulness for firewood, (4) usefulness for timber, (5) palatability of fodder, and (6) canopy effects on crops. Ranking each of several items, such as the trees, with respect to several criteria creates a matrix. This tool helps to understand how farmers trade off one trait with another during the decision-making process of species selection. Matrices and physical objects such as seeds (to construct the matrices) are used to develop a scoring system to evaluate a complex series of problems.

Modern variety – The product of formal, institutional (including NGO) and scientific plant breeding applying modern techniques of selection and technology and resulting in mostly homogeneous varieties/cultivars.

NGO – A non-governmental organization is a non-profit group organized outside of institutionalized political structures to realize particular objectives (such as environment protection). NGO activities range from research, information distribution, training, local organization and community service to legal advocacy, lobbying for legislative change and civil disobedience. NGOs range in size from small groups within a particular community to huge membership groups with a national or international scope.

Participation – Participation means that people become the stakeholders and decision-makers. Participation is the essence of responsible stewardship of natural resources. People's participation requires organization, interaction, consensus-building, decision-making and conflict resolution. In the research process, four kinds of participation have been widely used, depending upon the role of local people: contractual, consultative, collaborative and collegiate participation.

Participatory Crop Improvement (PCI) – This term is used to define a participatory approach to crop improvement which encompasses two contrasting methodologies, PPB and PVS.

Participatory Plant Breeding (PPB) – PPB is a breeding process in which farmers and plant breeders jointly select cultivars from segregating materials under a target environment. Other forms of PPB may also include activities such as germplasm enhancement through pure line or mass selection. By definition, PPB has significant farmer participation but it also involves decentralization of the breeding process from research station to farmers' fields. PPB approaches thus draw upon the comparative advantages of both the formal and informal systems. In recent years, PPB also has been considered as a potential strategy for enhancing biodiversity and production. In this case it is important to distinguish PVS and PPB as two distinct processes.

Participatory research – A framework for research which involves farmers in various processes of research in order to identify, design, test and evaluate new technologies which are appropriate to the needs of small farmers.

Participatory Rural Appraisal (PRA) – A set of facilitative and participatory techniques developed in the late1980s by researchers and NGOs, to understand the socioeconomic conditions and environment of a community to build upon the capabilities of local people and to empower them in the process. Increasingly this is also used as a research technique to help rural people analyze problems and develop their own solutions. Research methods are similar to RRA but the main emphasis is on the researcher facilitating communities to analyze their situation.

Participatory Technology Development (PTD) – PTD is a framework that involves farmers in various processes of the research cycle in order to identify, design, test and evaluate new technologies which are appropriate to the needs of small-scale farmers.

Participatory Variety Selection (PVS) – PVS is the selection of fixed lines (including landraces) by farmers in their target environments using their own selection criteria. The PVS consists of four methodological steps: (1) situation analysis and identification of farmers' varietal needs, (2) search for suitable genetic materials, (3) farmers' experimentation of new crop varieties in their own fields and with their own management practices, and (4) wider dissemination of farmer-preferred crop varieties.

Pedigree selection – Selection procedure in a segregating population in which progenies of selected F_2 plants are reselected in succeeding generations until genetic purity is reached.

Phenotype – (1) physical or external appearance of an organism as contrasted with its genetic constitution (genotype); (2) a group of organisms with similar physical or external make-up.

Plant Breeders' Rights (PBR) – PBR a form of intellectual property law that grants a plant breeder's certificate to those who breed new plant varieties. Plant breeders' rights generally contain breeders' and research exemptions that allow non-commercial use of protected varieties. There are currently two international agreements governing PBR, both of them under UPOV, the International Convention for the Protection of New Plant Varieties.

Population – In genetics, a group of individuals that share a common genepool and which actually interbreed.

Preference ranking – Preference ranking is a participatory way of obtaining an insight into how farmers make their decisions by asking them to rank several alternatives from best to least liked. This is commonly used in PPB and PVS for setting breeding goals and identifying farmer-preferred crop varieties.

Pure line – A strain in which all members have descended by self-fertilization from a single homozygous individual. A pure line is a genetically homozygous genotype.

Rapid Biodiversity Assessment (RBA) – RBA is a qualitative analytical tool that uses participatory and rapid information collection techniques in assessing biodiversity of a location. RBA uses a number of participatory appraisal techniques to suit different levels of biodiversity assessment.

Rapid Rural Appraisal (RRA) – RRA is an informal approach used to help rural people analyze problems and develop their own solutions in a rural setting. It is a set of tools developed in the early 1980s by researchers at universities and international research centres to gather and interpret data in a short time. There are four principal classes of RRA: (1) exploratory RRA, (2) topical RRA, (3) participatory RRA, and (4) monitoring RRA. A number of CG centres and other research and development institutions developed exploratory RRA with various names such as the Exploratory Survey in CIMMYT (Collinson 1981), Informal Agricultural Survey in CIP (Rhoades 1979), the Sondeo at ICTA in Guatemala (Hildebrand 1981), Rapid Assessment Technique developed by Northesat Rainfed Agricultural Development Project in Thailand (Alton and Craig 1987), the *Samuhik Brahman* (Joint or combined trek) developed at Lumle and Pakhribas Agricultural Centres in Nepal (Mathema and Galt 1987) and Diagnosis and Design developed by ICRAF (Raintree and Young 1983).

Restoration – The return of an ecosystem or habitat to its original community structure, natural complements of species and natural functions. In the case of agricultural crops, the return of landraces to farmers from genebanks after they have lost them owing to various types of disasters is also called restoration.

Roadside drama – This is a powerful theatrical tool to raise public awareness on conservation and use of plant genetic resources. Local farmers take part in street drama to sensitize fellow farmers on the value of plant diversity using poetry and songs and dramatizing at local sociocultural activities in a rural setting. A high level of participation can be guaranteed in this kind of activity.

Rural poetry journey – The rural poetry journey has been used as a new public awareness tool to convey value and uses of local plant genetic resources among rural and urban populations through on-the-spot writing and recital of biodiversity-related poems. A combined team of both national and local poets and poetesses travels together in the rural areas, interacting with the farmers, observing on-farm biodiversity of landraces, and translating their feelings and experiences into beautiful poems and songs. The songs and poems written along the journey are recited to the community members, conveying the message of conservation and utilization of local landraces.

Seasonal calendar – This is a useful PRA tool to outline an entire agricultural season, the crop sequences grown and associated tasks. It can supply information on environmental factors, as well as management decisions, value systems and labour responsibilities. A calendar identifies activities that vary from month to month. This tool also helps to understand biological and socioeconomic problems and constraints caused by the agricultural events. It is an extended version of the crop calendars which have long been used by agriculturists to describe cropping pattern in relation to time. All the major changes that occur within the farming year can be represented. Rainfall pattern, period of hail occurrence, frost, wind, peak labour demand period, drought, etc. can be superimposed. The changes in prices also may be included.

Segregation – The separation of homologous chromosomes (and genes) from different parents at meiosis.

Selection – (1) Any process, natural or artificial, that permits an increase in the proportion of certain genotypes or groups of genotypes in succeeding generations; (2) a plant, line or variety that originated by a selection process.

Social mapping – This is also a participatory method in which local people are encouraged to draw drainage, trails and important landmarks such as health post, office, bridges and other infrastructures. Then, each household of the village is mapped, a number assigned for each household and the name of the farmer is recorded. This tool is essential for baseline and intensive data plot studies where such information is needed for sampling and randomization as well as future follow-up and monitoring. This tool is very good to build rapport with the community and resource mapping can be done along with social mapping.

Species – A group of organisms capable of interbreeding freely with each other but not with members of other species; in taxonomic classification, a subdivision of a genus; a group of closely related individuals descended from the same stock.

***Sui generis* legislation** – Literally 'of its own kind', that is, in a class alone. This refers to any unique form of intellectual property legislation specifically designed to meet certain needs. This allows a country to develop its own alternative legal system of IPR to protect its plant varieties or even plant genetic resources.

Sustainable development – In the context of this bulletin, development that meets the needs and aspirations of the current generations without compromising the ability to meet those of future generations.

Timelines – A participatory research tool which focuses on charting events from the past, on facilitating the systemization of remembrance of events and chronologies of events.

Tolerance – Ability of plants to survive in the presence of a destructive pathogen, insect or environment condition.

Trade Related Intellectual Property Rights (TRIPS) – Trips is a GATT agreement, now administered by the World Trade Organization (WTO), stipulating that all signatories must conform to developed-country standards of intellectual property law. TRIPS requires signatories to introduce patent coverage for microorganisms and to have some form of intellectual property coverage for plants and animals. Developing countries have until at least the year 2000 to implement the agreement's IP provisions. Least-developed countries have to follow by 2004.

Transect map – Visual diagram constructed after a transect walk. It includes landscape features and associated problems, such as patterns of land use, technologies, environmental factors, species/varietal diversity and ethnicity which can be used for diagramming and mapping.

Transect walk – A systematic walk with a small group of farmers on their land to identify land-use system, problems and opportunities. It has particularly important role in overcoming the "roadside bias" that is common in field visits. A transect map can be drawn after a transect walk to assess biodiversity and the farming situation rapidly.

UPOV – Union for the Protection of new Varieties of Plants – a Geneva-based organization established under the World Intellectual Property Organization in 1961 to deal with plant breeders' rights. It has 30 members and seven others have initiated proceedings to join. There are two operative UPOV Conventions, dated 1978 and 1991. The 1978 Convention allows farmers to save and replant PBR-protected seed from their harvest. The 1991 version restricts the right of farmers to save seed and makes plant breeders' rights more like patents, extending the scope of the monopoly granted to the certificate holder. As of 5 January 1996, Australia, Denmark, Israel and Slovakia had ratified the more restrictive 1991 Convention. The UPOV Council meets every October, after a series of intergovernmental and government/industry committee meetings that regulate the Conventions' evolution. Many countries of the South are preparing to join UPOV.

Variety – In classical botany, a variety is a subdivision of a species. An agricultural variety is a group of similar plants that by structural features and performance can be identified from other varieties within the same species. The latter is synonymous with cultivar. In UPOV terms, a variety needs to be distinct, uniform and stable (DUS) to be protectable.

Village workshop – An effective method to assess usefulness of any intervention or forum to obtain feedback on ongoing activities or providing any information in a rural setting.

Wild relative – A non-cultivated species which is more or less related to a crop species (usually in the same genus) and has genetically contributed to the genome of the cultivated species.

World Trade Organization – A body created at the conclusion of the Uruguay Round of GATT in 1994 to monitor the GATT agreement and pursue global trade objectives. It became operational on 1 January 1996. It now has the potential to become the dominant forum for determining the future of intellectual property laws worldwide.

Bibliography

Introduction

CIAT. 1997. New Frontiers in Participatory Research and Gender Analysis. CGIAR Systemwide Program on Participatory Research and Gender Analysis for Technology Development and Institutional Innovation. CIAT, Cali, Colombia.

Eyzaguirre, P. and M. Iwanaga, eds. 1996. Participatory Plant Breeding. Proceedings of a workshop on participatory plant breeding, 26-29 July 1995, Wageningen, The Netherlands. International Plant Genetic Resources Institute, Rome.

McGuire, S., G. Manicad and L. Sperling. 1999. Technical and Institutional Issues in Participatory Plant Breeding - Do from a Perspective of Farmer Plant Breeding. A Global Analysis of Issues and Current Experience. CGIAR Systemwide Program on Participatory Research and Gender Analysis for Technology Development and Institutional Innovation. CIAT, Cali.

Sperling, L. and M. Loevinsohn, eds. 1996. Using Diversity: Enhancing and Maintaining Genetic Resources On-Farm. Proceedings of a workshop held on 19-21 June 1995, New Delhi, India. IDRC, New Delhi.

UPWARD. 1996. Into Action Research, Partnerships in Asian Rootcrop Research & Development. UPWARDS, Manila.

Veldhuizen, Laurens van, Ann Waters-Bayer, Ricardo Ramirez, Debra A. Johnson and John Thompson (eds.) 1997. Farmers' research in practice. ILEIA Readings in Sustainable Agriculture. Intermediate Technology Publications, London.

Section I. Crosscutting issues

Almekinders, C.J.M. and N.P. Louwaars. 1999. Farmers' seed production. New approaches and practices. Intermediate Technology Publications, London.

Amanor, K. 1990. Analyticial Abstracts on Farmer Pariticipatory Research. Agricultural Administration Unit Occasional Paper No. 9. ODI, London.

Ashby, J., C.A. Quiros and Y.M. Rivera. 1987. Farmer Participation in On-Farm Varietal Trials. Network Discussion Paper 22. Overseas Development Institute Agricultural Administration (Research and Extension), London.

Beebe, J. 1985. Rapid Rural Appraisal: The Critical First Step in a Farming Systems Approach to Research. Networking Paper No. 5. Farming Systems Support Project, Washington, D.C.

Biggs, S.D. 1989. Resource-Poor Farmer Participation in Research: A Synthesis of Experiences from Nine National Research Systems. OFCOR Comparative Study Paper No. 3. ISNAR, The Hague.

Boef, W. de, J. Hardon and N.P. Louwaars. 1997. Integrated organisation of institutional crop development as a system to maintain and stimulate the utilisation of agro-biodiversity at the farm level. Paper presented at the International meeting Managing Plant Genetic Resources in the African Savannah, Bamako Mali, 24-28 February 1997.

Byerlee, D. 1994. Modern varieties, productivity, and sustainability: recent experiences and emerging challenges. CIMMYT, Mexico.

Byerlee, D. and M. Collinson. 1980. Planning Technologies Appropriate to Farmers; Concepts and Procedures. CIMMYT Information Bulletin. CIMMYT, Mexico, D.F.

CGIAR, SGRP. 1997. New Frontiers in Participatory Research and Gender Analysis. Proceedings of the Internat-ional Seminar on Participatory Research and Gender Analysis for Technology Development, 9-14 September, 1996. CGIAR, Cali.

CGIAR, SWP/PRGA. 1997. New Frontiers in Participatory Research and Gender Analysis. Proceedings of the International Seminar on Participatory Research and Gender Analysis for Technology Development, 9-14 September 1996, Cali. CGIAR.

Chambers, R. 1990. Microenvironments Unobserved. Environment and Development Gatekeeper Series No. 22. IIED, London.

Chambers, R. and B. Ghildyal. 1985. Agricultural Research for Resource-Poor Farmers: the Farmer-First-and-Last Model. Discussion Paper No. 203. Institute of Development Studies, Brighton.

Chambers, R., A. Pacey and L.A. Thrupp, eds. 1989. Farmer First: Farmer Innovation and Agricultural Research. Intermediate Technology Publications, London.

CIAT. 1997. New Frontiers in Participatory Research and Gender Analysis. CGIAR Systemwide Program on Participatory Research and Gender Analysis for Technology Development and Institutional Innovation. CIAT, Cali, Colombia.

Clawson, D.L. 1985. Harvest security and intraspecific diversity in traditional tropical agriculture. Econ. Bot. 39:56-67.

Conway, G.R., J.A. McCracken and J.N. Pretty. 1987. Training Notes for Agro-ecosystem Analysis and Rapid Rural Appraisal. International Institute for Environment and Development, London.

Cornwall, A. and R. Jewkes. 1995. What is Participatory Research? Soc. Sci. Med. 41(12):1667-76.

Davis-Case, D. 1989. Community Forestry: Participatory Assessment, Monitoring and Evaluation. FAO Community Forestry Note 2. Food and Agriculture Organisation, Rome.

Davis-Case, D. 1990. The Community's Toolbox: The Idea, Methods and Tools for Participatory Assessment, Monitoring and Evaluation in Community Forestry. FAO Community Forestry Field Manual 2. Food and Agriculture Organisation, Rome.

De Boef, W, compiler. 1992. Local Knowledge and Agricultural Research: report on seminar, 28 September - 2 October 1992, Nyanga, Zimbabwe. CPRO-DLO/ Dept. of Sociology of Rural Development, Wageningen Agricultural University, Wageningen.

De Boef, W., K. Amanor, K. Wellard and A. Bebbington, eds. 1993. Cultivating Knowledge: Genetic Diversity, Farmer Experimentation and Crop Research. Intermediate Technology Publications, London.

Eyzaguirre, P. and M. Iwanaga, eds. 1996. Participatory Plant Breeding. Proceedings of a workshop on participatory plant breeding, 26-29 July 1995, Wageningen, The Netherlands. International Plant Genetic Resources Institute, Rome.

Eyzaguirre, P.B. and Ruth Raymond. 1995. Rural Women: A Key to the Conservation and Sustainable Use of Agricultural Biodiversity. Fourth World Conference on Women- Focus Day on Rural Women, FAO, Beijing

FAO. 1998. Socioeconomic and Gender Analysis (SEAGA): Field Handbook. Food and Agriculture Organization, Rome.

Farrington, J. and A. Martin. 1993. Farmer Participation in Agricultural Research: A Review of Concepts and Practices. Agricultural Administration Unit Occasional Paper no. 9. Overseas Development Institute, London.

Feldstein, H. and J. Jiggins (eds.) 1994. Tools for the Field: Methodologies Handbook for Gender Analysis in Agriculture. Kumarian Press, Connecticut.

Feldstein, H.S. and S.V. Poats, eds. 1989. Working Together. Gender Analysis in Agriculture. Vol. 2. Kumarian Press, West Hartford, Conn.

Fernández, M. and H. Salvatierra. 1989. Participatory Technology Validation in Highland Communities of Peru. Pp. 146-150 *in* Farmer First: farmer innovation and agricultural research (R. Chambers, A. Pacey and L. Thrupp, eds.). Intermediate Technology Publications, London.

Frankfort-Nachmias, C. and D. Nachmias. 1996. Research Methods in the Social Sciences. 5th edn. St. Martin's Press, New York.

Guarino, L. and E. Friis-Hansen. 1995. Collecting Plant Genetic Resources and Documenting Associated Indigenous Knowledge in the Field: A Participatory Approach. Pp. 345-346 *in* Collecting Plant Genetic Diversity: Technical Guidelines (L. Guarino, R. Rao and R. Reid, eds.). CAB International, Wallingford.

Guarino, L., R. Rao and R. Reid. 1995. Collecting Plant Genetic Diversity: Technical Guidelines. CAB International, Wallingford.

Guerrero, M. del P., J.A. Ashby and T. Garcia. 1993. Farmer evaluations of Technology: Preference ranking. Instructional Unit No. 2. CIAT, Cali.

Hannan-Andersson, C. 1992. Gender Planning Methodology: Three Papers on Incorporating the Gender Approach in Development Cooperation Programmes. Rapporteur Och Notiser 109. Institution for Kuffurgeografi och ekonomik geografi unid lund Universitet.

Hardon, J.J. and W.S. de Boef. 1993. Linking farmers and breeders in local crop development. Pp. 64-71 *in* Cultivating Knowledge. Genetic diversity, farmer experimentation and crop research (W. de Boef, K. Amanor, K. Wellard and A. Bebbington, eds.). Intermediate Technology Publications, London.

Hardon-Baars, A. 1997. User's Perspectives: Literature Review and Development of a Concept. UPWARD Working Paper Series No. 4. UPWARD, Los Banos, Laguna, Philippines.

Harrington, L. 1997. Doctors, lawyers and citizens: Farmer participation and research on natural resources management. Pp. 53-63 *in* New Frontiers in Participatory Research and Gender Analysis. CIAT, Cali, Colombia.

IDRC. 1998. Guidelines for Integrating Gender Analysis into Biodiversity Research. Sustainable Use of Biodiversity Program Initiative, IDRC, Ottawa.

Iriarte, Lucio, Litza Lazarte, Javier Franco y David Fernández. 1999. El Rol del Genero en la Conservación, Localización, y Manejo de la Diversidad Genética de Papa, Tarwi y Maiz. Gender and Genetic Resources Management. IPGRI, FAO, BIOSOMA, Rome.

Khon Kaen University. 1987. Proceedings of the 1985 International Conference on Rapid Rural Appraisal. Khon Kaen University, Thailand.

Lightfoot, C., C. Axinn, P. Singh, A. Bottrall and G. Conway. 1989. Training Resources Book for Agro-Ecosystem Mapping. IRRI, Phillipines.

Long, N. and A. Long (eds.). 1992. Battlefields of knowledge: the interlocking of theory and practice in social research and development. Routledge, London.

Lovelace, G., S. Subhadhira and S. Simaraks, eds. 1988. Rapid Rural Appraisal in Northeast Thailand: Case Studies. Khon Kaen University, Thailand.

Maxwell, S. 1984. Hitting a Moving Target; II. The Social Scientist in Farming Systems Research. Agriculture and Rural Problems Discussion Paper No. 199. Institute of Development Studies, Brighton.

McCracken, J., C. Kabutha and W. Ogana. 1992. Women, Conservation and Agriculture: A Manual for Trainers. Commonwealth Secretariat, London.

McCracken, J., J.N. Pretty and G.R. Conway. 1988. An Introduction to Rapid Rural Appraisal for Agricultural Development. International Institute for Environment and Development, London.

Molnar, A. 1989. Community Forestry: Rapid Appraisal. FAO Community Forestry Note 3. Food and Agriculture Organisation, Rome.

Mowbray, D. 1995. From Field to Lab and Back; Women in Rice Farming Systems. CGIAR, Washington, D.C.

Oakley, P. 1991. Projects with People: The Practice of Participation in Rural Development. ILO, Geneva.

Pimbert, M.P. and J.N. Pretty. 1995. Parks, people and professionals: putting "participation" into protected area management. UNRISD Discussion Paper no. 57. UNRISD, Geneva.

Prain, G. and J. Hagmann. 2000. Farmers' management of diversity in local systems. Pp. 94-100 *In* Encouraging Diversity. The synthesis between crop conservation and development (C.J.M. Almekinders and W.S. de Boef, eds.). Intermediate Technology Publications, London.

Prain, G., H. Fano and C. Fonseca. 1994. Involving Farmers In Crop Variety Evaluation and Selection. UPWARD Training Document Series 1994-2. UPWARD, Los Banos, Laguna.

Quiros, C.A., T. Garcia and J. Ashby. 1991. Farmer Evaluations of Technology: Methodology for Open-Ended Evaluation. Instructional Unit No. 1. IPRA Project/CIAT, Cali.

Sandoval-Nazarea, V. 1994. Memory Banking Protocol: A Guide for Documenting Indigenous Knowledge Associated with Traditional Crop Varieties. Training Document Series 1994-2. UPWARD, Los Banos, Laguna.

Schneider, J., ed. 1995. Indigenous Knowledge in Conservation of Crop Genetic Resources: Proceedings of an International Workshop Held in Cisarua, Bogor, Indonesia January 30 - February 3, 1995. CIP-EASEAP/CRIFC, Bogor.

Schönhuth, M. and Z. Kievelitz. 1994. Participatory Learning Approaches: Rapid Rural Appraisal, Participatory Appraisal, An Introductory Guide. TZ-Verlagsgesellschaft, RoBdorf.

Scoones, I. and J. Thompson. 1994. Beyond Farmer First: Rural People's Knowledge, Agricultural Research and Extension Practice. Intermediate Technology Publications, London.

Sen, A. 1989. Development as capability expansion. J. Development Planning No. 19.

Shrestha, P.K. 1998. Gene, Gender and Generation: Role of Traditional Seed Supply Systems in the Maintenance of Agrobiodiversity in Nepal. Pp. 143-152 in Managing Agrobiodiversity: Farmers' changing perspectives and institutional responses in the Hindu Kush-Himalayan region (Tej Partap and B. Sthapit, eds.). ICIMOD, IPGRI, Kathmandu.

Slim, H. and P. Thompson, eds. 1993. Listening for A Change: Oral Testimony and Development. Panos Panos Oral Testimony Program. Panos Publications Ltd., London.

Sperling, L. and P. Berkowitz. 1994. Partners in Selection: Bean Breeders and Women Bean Experts in Rwanda. CGIAR, Washington, D.C.

Subedi, A. and C. Garforth. 1996. Gender, Information and Communication Network: Implication for Extension. Eur. J. Agric. Educ. & Extension 32(2):63-74.

Tapia, Mario E. and Ana De la Torre. 1998. Women Farmers and Andean Seeds. Gender and Genetic Resources Management. IPGRI/FAO, Rome.

Tisch, S.J. 1994. Adressing Sustainability and Gender Issues in Asian Farming Systems Research. J. Asian Farming Systems Assoc. Vol. 2. 2.

Triomphe, B. 1995. Agroecología del Sistema de Aboneras en el Litoral Atlántico de Honduras. Paper presented at the Manejo Productivo y Sostenible de las Laderas, XLI Reunión Anual del PCCMCA, Tegucigalpa, Honduras, 27 al 31 de marzo 1995.

Uphoff, N.T. 1992. Local institutions and participation for sustainable development. IIED Gatekeeper Series no. 31. IIED, London.

Uphoff, N.T., J.M. Cohen and A.A. Goldsmith. 1979. Feasibility and application of rural development participation: a state-of-the-art paper. Cornell University, New York.

Warren, D.M. and K. Cashman. 1988. Indigenous Knowledge for Sustainable Agriculture and Rural Development. Gatekeeper Series No. SA10: Briefing papers on key sustainability issues in agricultural development. International Institute for Environment and Development, London.

Western, D. and M. Wright, eds. 1994. Natural Connections: Perspectives in Community-based Conservation. Island Press, Washington, D.C.

White, Sara C. 1996. Depoliticising development: The uses and abuses of participation. Development in Practice, Vol. 6, No. 1. Oxfam, UK and Ireland.

Wilde, V. and A. Vainio-Mattila. 1995. Gender Analysis and Forestry: International Training Package. FAO, Rome.

Section II. Enhancing farmers' access to plant genetic resources maintained ex situ

Ataro Adare and Bayush Tsegaye. 1994. Survey on relative performance of landraces for the 1993/94 cropping season. Consultancy Report, Seeds of Survival, USC/Canada in Ethiopia.

Bennett, E. 1970. Tactics of plant exploration. Pp. 157-129 *in* Genetic Resources in Plants: Their Exploration and Conservation (O.H. Frankel and E. Bennett, eds.). IBP Handbook No. 11. Blackwell Scientific Publishers, Oxford, UK.

Friis-Hansen, E. and D. Kiambi. 1998. Re-Introduction of Germplasm for Diversification of the Genetic Base of the Somali Agricultural Production System, IPGRI Technical Project Termination Report. IPGRI, Rome.

Friis-Hansen, E. and Rohrbach. 1993. SADC/ICRISAT 1992 Emergency Production of Sorghum and Pearl Millet Seed. Impact Assessment. ICRISAT Southern and Eastern Africa Region Working Paper 93/01. ICRISAT, Hyderabad.

Guarino, L. and E. Friis-Hansen. 1995. Collecting plant genetic resources and documenting associated indigenous knowledge in the field: a participatory approach. Pp. 345-361 *in* Collecting Plant Genetic Diversity (L. Guarino, V.R. Rao and R. Reid, eds.). CAB International, Oxon, UK.

Hodgkin, T. and M. Anishetty. 1999. Plant genetic resources and seed relief. Pp. 139-146 *in* Restoring Farmers' Seed Systems in Disaster Situations. Proceedings of the International Workshop on Developing Institutional Agreements and Capacity to Assist Farmers in Disaster Situations to Restore Agricultural Systems and Seed Security Activities, Rome, Italy, 3-5 Nov. 1998. FAO, Rome.

Nazarea-Sandoval, V.N. 1990. Potentials and limitations of ethnoscientific methods in agricultural reserarch. *In* Incountry Training Workshop for Farm Household Diagnostic Skills (R.E. Rhodes, V.N. Sandoval and C.P. Bagalanon, eds.). CIP, Los Banos.

Pistorius, R. 1997. Scientists, Plants and Politics – A History of the Plant Genetic Resources Movement. International Plant Genetic Resources Institute, Rome, Italy

Richards, P. and G. Ruivenkamp. 1997. Seeds and survival: Crop genetic resour-ces in war and reconstruction in Africa. IPGRI, Rome.

Scowcroft, W.R. and C.E.P. Scowcroft. 1997. Seeds security: Disaster response and strategic planning. Workshop paper. FAO, Rome.

Sperling, L. 1997. War and Crop Diversity. AGREN Network Paper No. 75. ODI, London.

Teshome, A., B. Baum, L. Fahrig, J.K. Torrance, J.T. Arnason and J.D. Lambert. 1997. Sorghum landrace variation and classification in north Shewa and south, Ethiopia. Euphytica 97:255-263.

Teshome, A., J.T. Arnason, J.K. Torrance, J.D. Lambert, L. Fahrig and B. Baum. 1999a. Traditional farmers' knowledge of sorghum landrace storability in Ethiopia. Econ. Bot. 53(1):69-78.

Teshome, A., L. Fahrig, J.K. Torrance, T.J. Arnason, J.D. Lambert and B. Baum. 1999b. Maintenance of sorghum landrace diversity by farmers' selection in Ethiopia. Econ. Bot. 53(1):79-88.

Teshome, Awegechew. 1996. Factors maintaining sorghum landrace diversity in north Shewa and south Welo regions of Ethiopia. PhD Thesis. Ottawa-Carleton Institute of Biology, Biology Department, Ottawa, Canada.

Tessema, Tesfaye. 1987. Improvement of indigenous wheat landraces in Ethiopia. Pp. 232-238 *in* Proceedings of the International Symposium on Conservation and Utilization of Ethiopian Germplasm, 13 –16 Oct. 1986 (J.M.M. Engels, ed.).

Vavilov, N.I. 1926. Studies on the origin of cultivated plants. Russian and English, State press, Leningrad. 248.

Vavilov, N.I. 1951. The origin, variation, immunity, and breeding of cultivated plants. *In* Selected writings of N.I. Vavilov. Translation by K. Starr, Chester. Chronica Botanica 13(1/16).

Worede, Melaku. 1992. Ethiopia: A gene bank working with farmers. Pp. 78-94. *in* Growing Diversity (D. Cooper, Renee Vellve and Henk Hobbelink, eds.). Intermediate Technology Publications, London.

Worede, M. and H. Mekbib. 1993. Linking genetic resources conservation to farmers in Ethiopia. Pp. 78-84 *in* Cultivating Knowledge: Genetic diversity, farmer experimentation and crop research (Walter de Boef *et al.*, eds.). Intermediate Technology Publications, London.

Section III. Local plant genetic resources management and crop improvement

Altieri, M.A. and L.C. Merrick. 1987. *In situ* conservation of crop genetic resources through maintenance of traditional farming systems. Econ. Bot. 41:86-96.

Baniya, B.K., A. Subedi, R.B. Rana, B.R. Sthapit, D.K. Rijal, C.L. Poudel and S.P. Khatiwada. 1999. Informal Rice Seed Supply and Storage Systems in Mid-hill of Nepal. Paper prepared for "The Third Global Participants Meeting on Strengthening the Scientific Basis of *In-situ* Conservation of Agricultural Biodiversity on-farm: 5-12 July 1999. Pokhara, Nepal.

Bellón, M.B. 1996. On-farm conservation as a process: An analysis of its components. Pp. 9-22 *in* Using Diversity: Enhancing and Maintaining Genetic Resources On-Farm (L. Sperling and M. Loevinsohn, eds). Proceedings of a workshop held on 19-21 June 1995, New Delhi, India. IDRC, New Delhi.

Berg, T., A. Bjornstad, C. Fowler and T. Skroppa. 1991. Technology options and the gene struggle. A report to the Norwegian Research Council for Science and Humanities. Development and Environment No. 8, NORAGRIC Occas-ional Paper Series C. Agricultural University of Norway.

Biggs, S.D. 1989. Resource-poor farmer participation in research: a synthesis of experiences from nine national agricultural research systems. OFCOR-comparative study paper no. 3. Special series on the organisation and management of on-farm client oriented research (OFCOR). International Service for National Agricultural Research (ISNAR), The Hague.

Brush, S.B. 1991. A farmer-based approach to conserving crop germplasm. Econ. Bot. 45(2):153-165.

Burt, R.S. 1980. Models of network structure. Ann. Rev. Soc. 6:79-141.

Castillo, F., L.M. Arias, R. Ortega and F. Marquez. 1999. Adding benefits through PPB, seed networks and grassroots strengthening. Proceedings of "The Third Global Participants Workshop" to develop tools and procedures for *in situ* conservation on-farm, 5-12 July 1999, Pokhara, Nepal

CBDC. 1996. Annual CBDC project report of 1996.

CBDC. 1997. Annual CBDC project report of 1997.

CBDC. 1998. Annual CBDC project report of 1998.

Ceccarelli, S., S. Grando and R.H. Booth. 1996. International breeding programme and resource-poor farmers: crop improvement in difficult environments. Pp. 99-116 *in* Participatory Plant Breeding (P. Eyzaguirre and M. Iwanaga, eds). IPGRI, Rome.

Chemjong, P.B., B.H. Baral, K.C. Thakuri, P.R. Neupane, R.K. Neupane and M.P. Upadhaya. 1995. The impact of Pakhribas Agricultural Centre in the Eastern Hills of Nepal: farmer adoption of nine agricultural technologies. Pakhribas Agricultiural Centre, Dhankuta, Nepal.

COWI. 1999. Seed sector study, Vietnam. Volume 1: Findings, Conclusions and Recommendations. Danida/MARD draft report. March 1999. Hanoi, Vietnam

De, Nguyen Ngoc and Huynh Quang Tin. 1998. Participatory plant breeding in Vietnam CBDC project. Technical report to CBDC Regional Co-ordination Unit.

De, Nguyen Ngoc. 1997. Data collection and analysis in the Mekong Delta Community Biodiversity Development and Conservation Project of Vietnam. Pp. 29 *in* Strengthening the scientific basis of *in situ* conservation of agricultural biodiversity on-farm: Options for data collecting and analysis (D. Jarvis and T. Hodgkin, eds.). Proceedings of a work-shop to develop tools and procedures for *in situ* conservation on-farm, 25-29 August 1997, Rome, Italy.

DTZ Peida. 1998. An evaluation study of participatory crop improvement in Nepal. A final Report. DTZ Peida Consulting, Edinburgh, UK.

Eyzaguirre, P. and M. Iwanaga. 1996. Farmers' contribution to maintaining genetic diversity in crops, and its role within the total genetic resources system. Pp. 9-18. *in* Participatory Plant Breeding (P. Eyzaguirre and M. Iwanaga, eds). Proceedings of a workshop on participatory plant breeding, 26-29 July 1995, Wageningen, The Netherlands. IDRC /FAO /CGN/ IPGRI, IPGRI, Rome, Italy.

FAO. 1996. The state of the World's Plant Genetic Resources for Food and Agriculture. FAO, Rome.

Friis-Hansen, E. 1999. The Socio-Economic Dynamics of Farmers' Local Plant Genetic Resources. A Framework for Analysis with Examples from a Tanzanian Case Study. CDR Working Paper 99.3. Copenhagen: Centre for Development Research. The publication can freely be downloaded from the internet: http://www.cdr.dk

Galt, D. 1989. Joining FSR to Commodity Programme Breeding Efforts earlier: Increasing Plant Breeding Efficiency in Nepal. Agricultural Administration (research and extension) Network: Network Paper 8. Overseas Development Institute, London.

Guarino, L. and E. Friis-Hansen. 1995. Collecting plant genetic resources and documenting associated indigenous knowledge in the field: a participatory approach. Pp. 345-365 *in* Collecting Plant Diversity: Technical Guidelines (L. Guarino, V. R. Rao, and R. Reid, eds.). IPGRI, FAO, UNEP, IUCN and CAB International, UK.

Hardon, J. and W.S. de Boef. 1993. Linking farmers and breeders in local crop development pp. 64-71 *in* Cultivating Knowledge Genetic Diversity, Farmer Experimentation and Crop Research (W.S. De Boef, K. Amanor, K. Wellard and A. Bebbington, eds.). ITP, London.

Hobbs, P.R., L.W. Harrington, C. Adhikari, G.S. Giri, S.R. Upadhyay and B. Adhikari. 1996. Wheat and Rice in the Nepal Terai: Farm Resources and Production Practices in Rupandehi District, NARC and CIMMYT.

Iglesias, C. and L.A. Hernandez R. 1996. Methodology development issues for participatory plant breeding of root and tuber crops. Pp. 129-134 *in* New Frontiers in Participatory Research and Gender Analysis. Proc. of the International Seminar on Participatory research and Gender Analysis for technology Development, Sept 9-14 1996. CGIAR Program on PRGA, CIAT.

Jaiswal, J.P., A. Subedi and K.J. Gurung. 1992. Seed production on cereals, grain legumes, oilseed and potato crops in LARC's command area: Existing system and review of the past work. Review Paper No. 1992/5. Lumle Regional Agricultural Research Centre, Pokhara, Nepal.

Jarvis, D. and T. Hodgkin. 1998. Farmer decision-making and genetic diversity: linking multidisciplinary research to implementation on-farm. *In* On-farm conservation: Issues and Case Studies (S. Brush, ed.). (in press)

Jarvis, D., T. Hodgkin, P. Eyzaguirre, G. Ayad, B.R. Sthapit and L. Guarino. 1998. Farmer selection, natural selection and crop genetic diversity: the need for a basic dataset. Pp.1-8 *in* Strengthening the scientific basis of *in situ* conservation of agricultural biodiversity on-farm. Options for data collecting and analysis (D. Jarvis and T. Hodgkin, eds.). Proceedings of a workshop to develop tools and procedures for *in situ* conservation on-farm, 25-29 August 1997, Rome, Italy.

Joshi, A. and J.R. Witcombe. 1996. Farmer participatory crop improvement. II: Participatory varietal selection, a case study in India. Exp. Agric. 32:461-477.

Joshi, K.D. and B.R. Sthapit. 1990. Informal Research and Development (IRD): A new approach to research and extension. LARC Discussion Paper 1990/4. Lumle Agricultural Research Centre, Pokhara, Nepal.

Joshi, K.D., B.R. Sthapit, R.B. Gurung, M.B. Gurung and J.R. Witcombe. 1997. Machhapuchre-3 (MP3), the first rice variety developed through a participatory plant breeding approach released for mid to high altitudes of Nepal. IRRN 1997:12.

Joshi, K.D., M. Subedi, R.B. Rana, K.B. Kadayat and B.R. Sthapit. 1997b. Enhancing on-farm varietal diversity through participatory varietal selection: a case study for *Chaite* rice in Nepal. Exp. Agric. 33:335-344.

Joshi, K.D., R.B. Rana, B. Gadal and J.R. Witcombe. 1998. The success of participatory varietal selection for Chaite rice in high potential production systems of in the Nepal Terai. *In* Proceedings of International conference on Food Security and Crop Science, Hissar, India, 3-6 November 1998. Hisar Agricultural University, Hisar, India (in press).

Kornegay, J., J.A. Beltran and J. Ashby. 1996. Farmer selections within segregating populations of common bean in Colombia. Pp. 151-160 *in* Participatory Plant Breeding (P. Eyzaguirre and M. Iwanaga, eds.). Proceedings of a work-shop on participatory plant breeding, 26-29 July 1995, Wageningen, The Netherlands. IDRC /FAO /CGN/ IPGRI.

LARC. 1995. The adoption and diffusion and incremental benefits of fifteen technol-ogies for crops, horticulture, livestock and forestry in the Western Hills of Nepal. LARC Occasional Paper 95/1. Lumle Agricultural Research Centre, Pokhara, Nepal. pp 69.

LARC. 1996. The adoption and diffusion and incremental benefits of fifteen technol-ogies for crops, horticulture, livestock and forestry in the Western Hills of Nepal. LARC Occasional Paper 97/1. Lumle Agricultural Research Centre, Pokhara, Nepal. pp. 30.

Maurya, D.M., A. Bottarall and J. Farrington. 1988. Improved livelihoods, genetic diversity and farmer participation: a strategy for rice breeding in rainfed areas of India. Exp. Agric. 24:311-320.

PRGA. 1999. Crossing perspectives. Farmers and scientists in participatory plant breeding. CGIAR Program on Partici-patory research and Genetic Analysis. CIAT, Cali.

Richards, P. 1996. Farmer knowledge and plant genetic resources management. Pp. 52-93 *in* *In situ* conservation and sustainable use of plant genetic resources for food and agriculture in developing countries (J. Engels, ed.). IPGRI, Rome, Italy.

Salazar, R. 1992. MASIPAG: alternative community rice breeding in the Philippines. Appropriate Technol. 18:20-21.

Salvaliya, T.B. 1997. New variety of groundnut. *In* International Conference on Creativity and Innovation at grassroots for Sustainable Natural Resource Management, 11-14 January 1997. Centre for Management in Agriculture, Indian Institute of Management, Ahmedabad, India.

Seta, R.B. 1997. Groundnut breeder. *In* International Conference on Creativity and Innovation at Grassroots for Sustainable Natural Resource Management, 11-14 January 1997. Centre for Management in Agriculture, Indian Institute of Management, Ahmedabad, India.

Soleri, D., S. Smith and D. Cleveland. 1999. Evaluating the potential for farmer-breeder collaboration: A case study of farmer maize selection from Oaxaca, Mexico. AgREN Network Paper No. 96a. ODI Agricultural Research and Extension Network, July 1999.

Sperling, L., M. E. Loevinsohn and B. Ntabomvra. 1993. Rethinking the farmer's role in plant breeding: Local bean experts and on-station selection in Rwanda. Exp. Agric. 29:509-519.

Sthapit, B.R. 1995. Variety testing, selection, and release system for rice and wheat crops in Nepal. Seed Regulatory Frame Works: Nepal. ODI/UK, CAZS/Lumle Agricultural Research Centre, Pokhara, Nepal.

Sthapit, B.R. and A. Subedi. 1999. Participatory variety selection and participatory plant breeding: An NGO experience and insights to support the local genetic resource base. *In* Encouraging Diversity. The Synthesis of Crop Conservation and Improvement (W.S. de Boef and C.J.M. Almekinders, eds.). IT Publications, London

Sthapit, B.R., K.D. Joshi and J.R. Witcombe. 1996. Farmer participatory crop improvement. III. Participatory plant breeding, a case study for rice in Nepal. Exp. Agric. 32:479-496.

Sthapit, B.R., K.D. Joshi and K.D. Subedi. 1994. Consolidating farmers role in plant breeding: A proposal for developing cold tolerant rice varieties for the hills of Nepal. LARC Discussion paper No. 94/ Pokhara, Nepal: Lumle Agricultural Centre.

Sthapit, B.R., K.D. Joshi, R.B. Rana and A. Subedi. 1998. Spread of varieties from participatory plant breeding in high altitude villages of Nepal. Local Initiatives for Biodiversity, Research and Development, Kaski, Pokhara, Nepal.

Subedi, A. and C. Garforth. 1996. Gender, information and communication net-work: implication for extension. Eur. J. Agric. Educ. & Extension 32(2):63-74.

Subedi, A., R.B. Rana and K.D. Joshi. 1997. Methodological approach to PPB: Experience from Nepal. Pp. 21 in Strengthening the scientific basis of *in situ* conservation of agricultural biodiversity on-farm: Options for data collecting and analysis (D. Jarvis and T. Hodgkin, eds.). Proceedings of a workshop to develop tools and procedures for *in situ* conservation on-farm, 25-29 August 1997, Rome, Italy.

Subedi, M. and P.K. Shrestha. 1999. Site selection report of Farmer-led Partici-patory Maize Breeding Programme for the Middle Hills of Nepal, 1998. LI-BIRD, Pokhara, Nepal.

Thakur, R. 1995. Prioritization and development of breeding strategies for rainfed lowlands: a critical appraisal. In Proc. of the IRRI conference "Fragile Lives in Fragile Ecosystems" IRRI, Los Baños, Philippines.

Vo-Tong Xuan. 1993. Present status of Agricultural Extension in Vietnam. Paper presented at the first Southeast Asia workshop on formulation of project proposals on technology transfer for major food crop production, FAO and UAF, HoChi Minh city, 6-9 Dec. 1993.

Weltzein/Smith, E., L.S. Meitzner and L. Sperling. 1999. Technical and Institut-ional Issues in Participatory Plant Breeding-Done from a Perspective of Formal Plant Breeding: A Global analysis of issues, results and current experience. Working Document No. 3, October 1999. CGIAR, Systemwide Program on PRGA for Technology Development and Institutional Innovation

Witcombe, J. 1999. Do farmer-participatory methods apply more to high potential areas than to marginal ones? Outlook on Agriculture 28(1):43-49.

Witcombe, J.R. 1997. Participatory approaches to plant breeding and selection. Biotechnology and Development Monitor No. 29, December 1996.

Witcombe, J.R. 1999. Participatory Plant Breeding and Broadening the Genetic Base of Crops. *In* Broadening the Genetic Base of Crops (D. Cooper, T. Hodgkin and C. Spillane, eds.). CAB International, UK. (in press)

Witcombe, J.R., A. Joshi, K.D. Joshi and B.R. Sthapit. 1996. Farmer participatory culti-var improvement. I: varietal selection and breeding methods and their impact on biodiversity. Exp. Agric. 32:445-460.

Section IV. Participatory approaches for establishing community seed banks and improving local seed systems

Aguirre, G., J. Calderon, D. Buitrago, V. Iriarte, J. Ramos, J. Blajos, G. Thiele and A. Devaux. 1999. Rustic Seedbeds, a Potential Bridge Between Formal and Traditional Potato Seed Systems in Bolivia. CIP Program Report 1996-98. CIP, Lima, Peru.

Cromwell, E., E. Friis-Hansen and M. Turner. 1992. The Organisation of the Seed Sector in Developing Countries, a Conceptual Framework of Analysis, pp. 1-81. ODI, London.

FAO. 1996. Global Plan of Action for the Conservation and Sustainable Utilisation of Plant Genetic Resources for Food and Agriculture and the Leipzig Declaration, pp. 3-63. International Technical Conference on Plant Genetic Resources, FAO, Rome.

Feyissa, R. 1997. *In situ* conservation of food crops: an Ethiopian model. *In* Proceedings of the workshop on planning and priority setting in eco-geographic survey and ethnobotanical research in relation to genetic resources in Ethiopia, 15-16 Feb. 1997, Addis Abeba.

Feyissa, R. 1998. A Dynamic Farmer-Based Approach to the Conservation of Ethiopia's Plant Genetic Resources. Project Progress Report, Addis Abeba.

Feyissa, R. 1999. Mainstreaming Biodiversity Conservation towards Sustainable Agricultural Development: An Ethiopian Perspective (In press).

Iriarte, V., A. Badani, C. Villarroel, G. Aguirre and E. Fernández-Northcote. 1998. Priorización, limpieza viral, producción de semilla de calidad básica y devolución de cultivares nativos libres de virus. PROINPA, Cochabamba.

PROINPA. 1998. Memoria. Primer Encuentro

Taller sobre el Mantenimiento de la Diversidad de Tuberculos Andinos en su Zonas de Origen. PROINPA, Cochabamba.

Scheidegger, U., G. Prain, F. Ezeta and C. Vittorelli. 1989. Linking formal R&D to indigenous systems: a user oriented seed programme for Peru. ODI, London.

Tapia, M.E. and A. de la Torre. 1993. La mujer campesina y las semillas andinas. FAO, Lima, Peru.

Thiele, G. 1999. Informal potato seed systems in the Andes: Why are they important and what should we do with them? World Development 27(1):83-99.

Tripp, R., N. Louwaars, W. Joost van der Burg, D.S. Virk and J.R. Witcombe. 1997. Alternatives for Seed Regulatory Reform, an Analysis of Variety Testing, Variety Regulation and Seed Quality Control, pp. 1-25. Agricultural Research & Extension Service, ODI, London.

Wiggins, S. and E. Cromwell. 1995. NGOs and seed provision to smallholders in developing countries. World Development 23(3):413-422.

Worede, M. 1992. Ethiopia: A genebank working with farmers. In Growing Diversity. Genetic Resources and Local Food Security (D. Cooper, Renee Vellve and Henk Hobbelink, eds.). Intermediate Technology Publications, London.

Worede, M. and H. Mekbib. 1993. Linking genetic resources conservation to farmers in Ethiopia. Pp. 78-84 in Cultivating Knowledge: Genetic Diversity, Farmer Experimentation and Crop Research (Walter de Boef, Kojo Amanor, Kate Wellard and Anthony Bebbington, eds.). Intermediate Technology Publications, London.

Section V. Increasing public and policy awareness of conservation and use of plant genetic resources

Gauchan, D., A. Subedi, S.N. Vaidya, M.P. Upadhaya, B.K. Baniya, D.K. Rijal and P. Chaudhari. 1999: National Policy and its Implication for Agrobiodiversity Conservation and utilization in Nepal. Paper presented in "Third Global Participant Meeting on *In situ* Agrobiodiversity Conservation On-farm" held in Pokhara, Nepal, July 5-12, 1999.

Gautam, J.C., J.N. Thapaliya and H. Bimb. 1999. Policy on Agrobiodiversity in Nepalese National Context. Paper presented in National Conference on Wild Relatives of Cultivated Plants in Nepal, June 2-4, 1999. Organized by Green Energy Mission/Nepal.

Gills, G. 1992. Policy analysis for agricultural resources management in Nepal. A comparison of conventional and partici-patory approaches. HMG/Winrock International, Kathmandu, Nepal.

Gills, G. 1998. Using PRA for agricultural policy analysis in Nepal. The Tarai Research Network Food Grain Study. Holland and Black Burry, eds. "Whose Voice" Participatory Research and Policy Change.

Jarvis, Devra I. and Toby Hodgkin, eds. 1998. Strengthening the Scientific Basis of *In situ* Conservation of Agricultural Biodiversity On-farm. Options for Data Collecting and Analysis. IPGRI, Rome, Italy.

Pant, R. 1998. Addressing Broader Issues of Biodiversity /Agrobiodiversity Conservation within the Framework of National Policy: A Case from Nepal. In Managing Agrobiodiversity: Farmers' Changing Perspectives and Institutional Responses in the Hindu Kush-Himalayan Region (Pratap and B.R. Sthapit, eds.). ICIMOD/IPGRI.

Rijal, D.K., R.B. Rana and K.D. Joshi. 1997. Conservation of Non-Renewable Plant Resources: Problems and Prospects of agrobiodiversity conservation in Nepal. Paper presented for the national seminar on the "The National Biodiversity Action Plan" held on 24th October, 1997, Organized by RESOURCES NEPAL, Kathmandu, Nepal.

Rijal, D.K., B.R. Sthapit, R.B. Rana, P.R. Tiwari, P. Chaudhary, Y.R. Pandey, C.L. Poudel and A. Subedi. 1999. Options for adding values and benefits of local crop diversity through a non-breeding approach: Experiences of Jumla, Kaski and Bara sites in Nepal. Paper presented at the Third Participants Meeting of the Project, "Strengthening the Scientific Basis for *in situ* conservation of Agriculture Biodiversity On-farm", co-hosted by NARC, LI-BIRD and IPGRI held at Pokhara, Nepal,

5-12 July 1999.

Robb, C. 1998. Participatory Policy Analysis (PPA): A review of World Bank's experience. Holland and Black Burry, eds. "Whose Voice" Participatory Research and Policy Change.

Sthapit, B.R., D. Gauchan and R.B. Rana. 1996. Rethinking in Research: Participatory Plant Breeding in Outreach Research Sites. *In* Proceeding of the Third National Workshop on Outreach Research, held on 23-25 May, 1996, Outreach Research Division, Nepal Agricultural Research Council, Khumaltar, Nepal.

Swaminathan, M.S. 1998. Farmers' Rights and Plant Genetic Resources. Biotechnol-ogy and Development Monitor No. 36. Various Options for *sui generis* rights systems. University of Amsterdam.

Tapia, M.E. and A. Rosa. 1993. Seed fairs in the Andes: a strategy for local conservation of plant genetic resources. Pp. 111-118 *in* Cultivating Knowledge: Genetic Diversity, Farmer Participation and Crop Research (Walter de Boef, K Amanor, K Wellard and A Bebbington, eds.). IT Publications, UK.

Tripp, R. and W. Heide. 1996. The erosion of crop genetic diversity: Challenges, strategies and uncertainties. ODI Natural Resources Perspectives, No. 7, March 1996. ODI, London.